MW01119773

THE NEW MEDIA NATION

Anthropology of Media
Series Editors: John Postill and Mark Peterson

The ubiquity of media across the globe has led to an explosion of interest in the ways people around the world use media as part of their everyday lives. This series addresses the need for works that describe and theorize multiple, emerging, and sometimes interconnected, media practices in the contemporary world. Interdisciplinary and inclusive, this series offers a forum for ethnographic methodologies, descriptions of non-Western media practices, explorations of transnational connectivity, and studies that link culture and practices across fields of media production and consumption.

Volume 1
Alarming Reports: Communicating Conflict in the Daily News
Andrew Arno

Volume 2
The New Media Nation: Indigenous Peoples and Global Communication
Valerie Alia

The New Media Nation

Indigenous Peoples and Global Communication

Valerie Alia

Berghahn Books
NEW YORK • OXFORD

First published in 2009 by

Berghahn Books

www.berghahnbooks.com

Library of Congress Cataloging-in-Publication Data

Alia, Valerie, 1942-
The new media nation : Indigenous peoples and global communication / Valerie Alia.
p. cm. – (Anthropology of the media)
Includes bibliographical references and index.
ISBN 978-1-84545-420-3 (hardback : alk. paper)
1. Indigenous peoples–Communication. 2. Indigenous peoples and mass media. 3. Internet and indigenous peoples. 4. Telecommunication. 5. Communication and culture. I. Title.
GN380.A45 2009
302.23089'97–dc22

2009025363

British Library Cataloguing in Publication Data

A catalogue record for this book is available from the British Library

Printed in the United States on acid-free paper.

ISBN: 978-1-84545-420-3 Hardback

To the world-changing pioneers
who are creating and sustaining The New Media Nation

Contents

Figures

Tables

Preface

This is the moment freedom begins—the moment you realize someone else has been writing your story and it's time you took the pen from his hand and started writing it yourself. The greatest challenge to the ... media giants is the innovation and expression made possible by the digital revolution. I may still prefer the newspaper for its investigative journalism and in-depth analysis, but we now have ... the means to tell a different story than big media tells.... This is the real gift of the digital revolution. The Internet, cell phones and digital cameras that can transmit images over the Internet, make possible a nation of story tellers ... it's not a top-down story anymore. Other folks are going to write the story from the ground up...
—Bill Moyers, US broadcaster and journalist

Native people are doing for themselves what cannot be accomplished by the mainstream media. They are sharing their communities' concerns in their own voices, uninterrupted by cultural interpreters and reporters who lack the background to understand the complex issues of contemporary Native life...
—Peggy Berryhill, Native American broadcast producer

To me one of the answers comes through Native people creating books, films and radio stations. I believe control of our own institutions is going to empower us; being able to get the story, tell our story, work and edit our stories ... For the first time Native people are on the breaking edge

> of information technology in terms of computer systems and the internet, which means that we're going back to an old tradition, the oral visual presentation and the storyteller's credibility.
> —Paul DeMain, Oneida/Ojibway journalist

Thus, Paul DeMain brings things full circle. Old traditions flow into new technologies; change is organic, not anomalous. DeMain is a leading Indigenous journalist of Oneida/Ojibway ancestry, a publications editor, radio presenter, and media activist who pointedly titles the piece from which that paragraph is quoted, "Guerillas in the Media." Guerillas, indeed. Like other Indigenous media leaders around the globe, DeMain is at once a 'guerilla' and a 'legitimate' journalist—not either...or, but both. That is a theme to which we will return throughout this exploration of The New Media Nation. In 2004, reflecting on the end of the United Nations Decade of Indigenous Peoples, I wrote:

> The world's Indigenous peoples have experienced both progress and frustration in the past decade. Important developments include a comprehensive Arctic Policy brought in by the ICC—Inuit Circumpolar Council (formerly Inuit Circumpolar Conference)—one of the first NGOs to gain UN recognition; signing of the Nunavut Agreement between Inuit and the Government of Canada and founding in 1999 of Nunavut Territory, whose 85% Inuit majority has helped to spark a cultural revival across the circumpolar world. It was preceded by Greenland's acquisition of home rule from Denmark. Siberian peoples are culturally thriving—and dying from economic, environmental and social conditions. Residential/missionary school abuse is variously addressed and ignored in Canada, Australia, the United States and other countries.
>
> The ICC's Arctic policy could not be timelier in the face of global warming, arguments over Kyoto and proposed industrialization of fragile lands and waters. Amazonian peoples and their environments remain threatened. On the proactive side, the Fair Trade movement is attempting to address environmental, social and economic problems on a global scale, and in the Nordic countries Sámi cultural, social, political and communications advances join the world-wide movement of Indigenous-run media networks that I have called "The New Media Nation." I think it heralds the rise of a new media universe. Until recently, representations of Indigenous peoples were primarily by outsiders and Indigenous people were excluded from positions of influence in media industries.They were portrayed as exotic items for study or observation, in need of "civilizing."

> Women—in the vanguard of many movements, including the "Māori Renaissance" and what might be called the Inuit cultural and political renaissance—were marginalized or trivialized (Alia 2004b).

Previous publications have illuminated particular facets of the Indigenous media picture. These include the pivotal studies of broadcasting in Canada by Lorna Roth and Gail Valaskakis (Roth 2006, 2000, 1996, 1995, 1993; Roth and Valaskakis 1989); Michael Meadows' and Helen Molnar's work on Australian Aboriginal and Torres Strait Islander media (Meadows 2001; Meadows and Molnar 2002; Molnar and Meadows 2001); Ritva Levo-Henriksson's (2007) study of Hopi radio; Ann Fienup-Riordan's (1995) study of Alaska Eskimos in film; the autobiographical writings of Inuit photographer/writer/artist Peter Pitseolak (Pitseolak 1997; Pitseolak and Eber 1975, 1993); the historically and politically grounded studies of Sámi media by Veli-Pekka Lehtola (2002, 2005) and Sari Pietikäinen (2003); Michael Keith's (1995) work on Native American broadcasting; work by Patrick Daley and Beverly James (2004) on Alaska Native media; Donald R. Browne's (1996)[1] study of electronic media; Neil Blair Christensen's (2003) study of Internet use in Greenland; collections on Indigenous literature in North America (Swann 2004; Harjo and Bird 1997; Petrone 1992, 1990, 1988); international collections focused mainly on film and video, (Stewart and Wilson 2008; Ginsburg, Abu-Lughad, and Larkin 2002) and on Africa (Wasserman and Jacobs 2003); collected essays on Aotearoa/New Zealand (Dennis and Bieringa 1996); my earlier Canada-based work (Alia 1999) and first look at the international perspective, in the essay, "Scattered Voices, Global Vision: Indigenous Peoples and the 'New Media Nation'" (Alia 2003).

Franz Fanon observed that, "Mastery of language affords remarkable power," because it results in possession of "the world expressed and implied by that language" (Fanon 1967: 18). UNESCO has designated Indigenous and global networks urgent areas for study and policy development; organizations such as Survival International and Cultural Survival have made media a top priority. In Indigenous communities, worldwide, education, broadcasting, publishing, and community-based programs are developing simultaneously (Tuhiwai Smith 1999: 147). This book looks at the role of Indigenous media in promoting such culturally specific revitalization initiatives, along with a broader pan-Indigenous agenda. This is the first overview of a powerful international movement that looks both inward and outward, simultaneously helping to preserve ancient languages and cultures and communicating across cultural, political, and geographical boundaries.

Acknowledgements

I thank the Canadian High Commission (UK) and Leeds Metropolitan University for funding research undertaken for this book; Markku Henriksson, Ritva Levo-Henriksson, Michael Meadows, Michael Posluns, and Lorna Roth for friendship, inspiration, and groundbreaking work, and Michael for generously sharing photographs; Yutaka Jose and Gabriele Hadl for illuminating Ainu concerns; Pamela Bruder, David Goldberg, and Cam McDonald for sending materials I had missed; Alan Brewis for explaining the importance of cell phones in Africa. For information, and provocative and illuminating discussions, I thank Keith Battarbee, David Bell, Jim Bell, Michael Bravo, Holly Bright, Debbie Brisebois, Rávdná Nilsdatter Buljo, Dumitru Chiţoran, Allan Chrisjohn, Helen Fallding, Frederick Fletcher, Carol Geddes, Kathryn Hazel, Nils Johan Heatta, Kaisa Rautio Helander, Ken Kane, Cheris Kramarae, Rob Milen, Miles Morrisseau, Ludger Müller-Wille, Juhani Nousuniemi, Lee Selleck, Lorna Roth, Scott Smith, Dan Smoke, Mary Lou Smoke, Marshall Soules, Sergio Urriaga, Bud Whiteye,[1] and Barry Zellen.

In Leeds, I thank Simon Lee for his 'running stream' vision; Gavin Fairbairn, Sheila Scraton, John Shutt, Hilary Sommerlad, Dave Webb, and Barry Winter for placing ethics, diversity, and community front and center; Seidu Alidu, Ayeray Medina Bustos, Mustafa Jamil Iqbal, and Lucy May Reynolds; Ritchard Emm and Emily Marshall for outstanding research assistance, and Ritchard for saintly patience with an unreconstructed technophobe. I thank Andrew Arno for his thoughtful and extremely helpful reading of the manuscript; Pete Steffens for eagle-eyed editorial expertise and translation of Russian texts; the wonderful team at Berghahn Books—Marion Berghahn for a multitude of insights and kindnesses, Ann Przyzycki, Melissa Spinelli, Katherine Elgart, and Cassandra Caswell; and, as always, the family—

Dan, Dave, Mary Margaret Zahara, Peggy Jane, Sivan, Daneet, Alan, Susan, Rachel, Bill, Margaret, Sylvia, and Pete, for life- and sanity-support and so much more. The two additional acknowledgements that follow regrettably can no longer be made in person.

In Memoriam: Conway Waniente Jocks and Gail Guthrie Valaskakis

In July 2007, the world lost two of the great pioneers of Indigenous broadcasting. I had the good fortune to work with both of them and call them friends, and thus, their loss is personal as well as professional.

The website of CKRK-FM, the station Conway founded, published a large, beautiful photograph of his face (CKRK-FM 2007). There was no long obituary, just a few powerfully moving words that said it all:

Conway Waniente Jocks
Founder of CKRK-FM
August 4 1927–July 5 2007
Without you, there would be no US.

Conway influenced my own thinking, not just on Indigenous media, but also on community radio and political cartooning. He was a renaissance person in the truest sense—an artist, writer, historian, cartoonist, broadcaster, and activist. In 1999, *The Eastern Door*, the Kahnawake newspaper for which he penned countless cartoons and articles, wrote:

> Conway's first editorial cartoon appeared in his high school newspaper in the 1940's... He claims that there were few, if any, conflicts in the editorial direction since he was also the editor. His first job, with the Chicago Tribune's New York office, was cut short when he was drafted into the US Army (Eastern Door 1999).

Describing his career, he said: "Early on, I evolved a personal, one-line code of ethics: No one tells me what I can or cannot draw." It was to cost him more than one job, including coveted work in New York on the Batman comic. After a stint in photogrammetry, he returned home to Kahnawake, adopted the pen name Tsiti and drew cartoons for magazines. He joined the Cartoonists' Guild of America, began work as Communications Coordinator for the newly formed Kanien'kehaka Raotitiohkwa Cultural Center, and designed and illustrated a series of booklets on Kahnawake history. In 1996 and 1997, he won awards for best editorial cartoon in Québec and in Canada. The distinguished car-

Conway Waniente Jocks at the CKRK Radio Hut, 1991 pow wow at Kahnawake Mohawk Nation (Canada). Photograph by Valerie Alia.

toonist, Terry Mosher, known as Aislin, praised Conway's "gentle but occasionally cutting style, adding, 'He's very sly'" (Eastern Door 1999).

Gail Guthrie Valaskakis was a gentle activist and scholar, known best for her work in Aboriginal broadcasting. She "liked to say that she was born 'with a moccasin on one foot and a shoe on the other"' (Hustak 2007). The 'moccasin' came from her Chippewa father, the 'shoe' from her Dutch-American mother. She was born at Lac du Flambeau, Wisconsin, attended a federal Indian school, "and all her life identified as aboriginal" (Hustak 2007). She was a founder of the Montreal Native Friendship Centre, Manitou Community College, Montreal's Native North American Studies Institute, and with her partner, Stan Cudek, the Waseskun Healing Centre. She earned a masters degree in theatre arts from Cornell University, moved to Canada in 1966 and taught at Kahnawake (Mohawk Territory, Québec) and Loyola College (later Concordia University), where she became a Professor of Communication Studies and Dean of Arts and Science. Her Ph.D. from McGill University addressed the impact of new media technologies on Aboriginal culture. The research took her to the Arctic, where she became far more than a scholar, playing a major role in the development of Indigenous media. "Three decades before the advent of the Aboriginal Peoples Television Network, she was helping [to] establish [the] first media and communications programs" (National Post 2007).

In 2006, Gail told me that it was her last job that she most loved. From 1998 until a few months before her untimely death in 2007, she was director of research for the Aboriginal Healing Foundation, responsible for managing the $350-million fund, which was established by Ottawa following the Royal Commission on Aboriginal Peoples to help Indigenous people cope with the effects of their experiences in residential schools. She "used her expertise to guide the 1,300 projects the Foundation has funded over the last decade... [and] bettered the lives of countless [people] with quiet good works" (National Post 2007). Georges Erasmus, former grand chief of Canada's Assembly of First Nations and president of the Aboriginal Healing Foundation, said, "I can't say enough about her. She was loved by all. Her death has left a

Gail Guthrie Valaskakis. Courtesy of Concordia University Archives, HA3446.

giant hole in the foundation" (Hustak 2007). Gail's long-time friend and colleague, Lorna Roth, called her "a memorable teacher, a fascinating cross-cultural scholar, one of those rare people who could fit into multiple places, who could communicate graciously and with great sensitivity to anyone from any walk of life" (Hustak 2007). A blogger called 'balbulican' wrote: "Gail helped build a movement, a body of knowledge, and a revolutionary new application of broadcasting and satellite technology as a tool for social development instead of assimilation" (balbulican 2007).

Photo Credits and Permissions to Use other Images and Illustrations

The Aboriginal Peoples Television Network

Valerie Alia

Robin Armour

Brett Ericksson, courtesy of *Fellowship* magazine (www.forusa.org/fellowship)

Karen Herland, Assistant Editor, *Concordia Journal*, Concordia University

The Inuit Broadcasting Corporation

Mustafa Maluka, courtesy of *Fellowship* magazine

Rachel Marion, Archivist, Concordia University Archives

Nancy Marrelli, Director, Concordia University Archives

Michael Meadows

Northern Native Broadcasting Yukon and Television Northern Canada

Sámi Radio Finland

Sámi Radio Norway

Francisco Vazquez, courtesy of *Fellowship* magazine

Ethan Vesely-Flad, Communications Co-ordinator and Editor of Fellowship magazine, The Fellowship for Reconciliation

Notes on Language and Research Methods

Language, Spelling, Translation, and Transliteration

Preferred usage varies among peoples and regions. Although many Indigenous organizations have veered away from *Native*, and many Native Americans in the US (and Indigenous people in other countries) find it offensive, it is still used widely in Canada—for example, in the name of Northern Native Broadcasting, Yukon (NNNBY).

In Canada, *Inuit* (singular *Inuk*, language *Inuktitut*) is preferred, except in reference to *Inuvialuit*—Inuit of the western NWT. *Inuit* is also used generically to refer to all of the members of the Inuit Circumpolar Council (ICC), although individually they often use other terms. Inuit Alaskans have preferred *Eskimo*, but increasingly choose the names of their distinctive cultures (e.g., *Yup'ik*; *Iñupiat*). Indigenous Greenlandic people also have separately named cultures, use the generic *Inuit* in relation to their participation in the global organization, Inuit Circumpolar Council, but usually call themselves *Kalaallit* to indicate their common allegiance to Kalaallit Nunaat (Greenland under home rule). The term, *First Nations*, "came into common use in the 1970s" (Assembly of First Nations 2009). Widely preferred by many Indigenous Canadians (and increasingly, by some Native Americans), it excludes Inuit and Métis, and therefore omits a major portion of indigenous northerners. Others prefer the more inclusive *First Peoples*, or *Aboriginal*, though in Europe, the UK, and elsewhere, it is often taken to mean Aboriginal Australian people. Within Australia, *Indigenous* is increasingly favored.

Where others are quoted, the whole range of terms appears, scattered throughout the book. I have given preference to the internationally accepted and more inclusive *Indigenous* or *First Peoples*. There is no wholly successful way to refer to people who are not Indigenous. *Non-Indigenous* is admittedly a dodge. Inuit call non-Inuit people *Qallunaat* (singular, *Qallunaaq*) (which does not refer exclusively to "white" people). Māori call outsiders Pākehā. There is no universal solution; the terms remain, literally, all over the map. I use the spelling Sámi; others use Saami or Sami. I refer to the Sámi homeland as Sápmi, because it is more widely used than the alternate, Sámiland. I capitalize Indigenous, while others do not, mainly because the lower-case "indigenous" is sometimes used by governments and others, to mean merely "local" or "national". I use the spelling, Māori, while some Māori and Pākehā prefer Maori or Mãori, and both *Aotearoa* (the Māori name) and *New Zealand*, depending on the context. I refer to Burma rather than Myanmar, because it is preferred by Indigenous and other peoples who challenge government brutality and protest the renaming of their country.

As for the unifying term on which this book revolves, the truth is that the very question of indigeneity remains open and unresolved. No one has come up with a universal definition of "Indigenous," and indeed, there is no universal agreement among the peoples who use this term to refer to themselves and others. The United Nations Working Group on Indigenous Populations refers to "pre-invasion and pre-colonial societies" that consider themselves coherent cultural groups and estimates "350 million individuals (four percent of humanity), representing over 5,000 languages and cultures in more than 70 countries." There is no internationally agreed definition of indigenous peoples and ILO Convention No. 169 makes self-identification the central identifying principle (International Labour Organisation 2005: Article 1). The Sámi Parliament concurs, defining an Indigenous people as one whose members consider themselves Indigenous, whose "ancestors have inhabited the area before it was conquered or settled or before the present borders were drawn up ... [and has] distinct social, economic, cultural and political institutions" (Sami Parliament 2002: 3). The Māori organizers of the 2007 global Indigenous media conference chose a representative of Welsh radio S4C, Chairman John Walter Jones, as one of the keynote speakers, though not everyone would see the Welsh as an Indigenous people (WITB 2007). According to the Sámi Parliament, "A people is considered *an indigenous people* if its ancestors have inhabited the area before it was conquered or settled or before the present borders were drawn up. In addition, such a people is to have their distinct social, economic, cultural and political institutions and to consider themselves an indigenous people" (Sami Parliament 2002: 3).

Research Methods

This is a unique study of an evolving, world-changing global media movement. I have employed several methods at different stages of more than twenty years of research, and have published portions of the work, along the way. The research was supported by grants from the American Association of University Women (AAUW) Educational Foundation, the Social Sciences and Humanities Research Council of Canada, the (US) National Science Foundation, the Canadian High Commission in the UK, Western Washington University, York University, the University of Western Ontario (UWO), and most recently, the University of Sunderland and Leeds Metropolitan University. Canada's Royal Commission on Electoral Reform and Party Financing (RCERPF) commissioned a study of northern Indigenous people and media; it included interviews, library and archival work, and review of hundreds of transcripts from testimony given to public hearings, which the Commission organized across Canada. The findings were reported at a meeting and in a chapter for one of the volumes of RCERPF studies, (Alia in Milen 1991: 105–146) and informed a series of recommendations submitted to the Commission. Other facets of the Canadian research were reported in my book, *Un/Covering the North* (Alia 1999).

I used questionnaires for a preliminary study of Indigenous media organizations; was a participant observer, an advocate and activist with several Indigenous media organizations and projects. The billeting and room sharing common across the North facilitated participant observation. I witnessed the process leading to closure of UWO's Programme in Journalism for Native People (PJNP), described in *Un/Covering the North,* and conducted several sets of interviews with Indigenous writers and broadcasters, local, provincial, territorial, and national politicians and government representatives, Indigenous and non-Indigenous community leaders, journalists, and audiences. Interviews were conducted by teams under my supervision, in Nunavut, the Northwest Territories, Yukon, and northern regions of the Canadian provinces. I hired researchers from different locations, cultures, and backgrounds and—as much as possible—asked them to gather data from their home communities. They included members of several First Nations in Yukon, the western Arctic and Northern provinces, Inuit from Nunavut, and non-Indigenous people from several provinces. I conducted a two-year study of print media coverage of the North with the help of student assistants from the University of Western Ontario (UWO) Graduate School of Journalism (among them the now well-known journalists, Adrienne

Arsenault (CBC Television) and Kjersti Strømmen (NRK, Norwegian television).

In the 1990s, I conducted a three-year case study of communications in Canada's Yukon Territory (Alia in Milen 1991: 105—146; Alia 1996). Most studies of the Canadian North focus on the Northwest Territories and Nunavut, with Yukon receiving scant attention, perhaps because of its small land mass and even smaller population—about 33,000 at the time of the study. The Yukon study provided a detailed view of northern communications and communicators, with a community-by-community survey of what is produced, in what manner, and by whom, who its audiences are, and how they receive and perceive it. The primary work was done in Whitehorse, where most government, private, and communications institutions are based, with frequent and extended travel to most Yukon communities. For each of the three years, I lived in Yukon during autumn and winter, when tourists and researchers seldom come. The experience brought home an old lesson about research methods: no matter how well you do your homework, nothing can replace first-hand experience. Some assumptions proved mistaken—for example, that I could understand how all of northern Canada experienced communications by extrapolating from previous work in the eastern Arctic, where there are no roads between communities, and travel is limited to air, snowmobile, all-terrain vehicle (ATV), *qamotiq* (Inuit sled), small watercraft, and supply ship. I had a respectable understanding of the time and expense involved in conducting northern travel, and wrongly imagined that access would be greater and less costly, because all Yukon communities except Old Crow are (theoretically!) accessible by road. I learned that costs are as high, and scheduling as complex, as in the relatively roadless Northwest Territories and Nunavut. Roads are often dangerous or impossible to negotiate. Bus routes link many communities; on paper, there are schedules. However, even when buses run on time, there is not always room for passengers. During winter holidays one year, the van from Whitehorse to Atlin, British Columbia was filled with holiday mail; there was space for just one passenger. I had fortunately booked early; others waited at least two days for seats.

In three years, I was able to visit most Yukon communities, but not all of them, even after budget and time were extended. Close observation of one region for an extended time afforded crucial insights into daily lives and work, and the ways in which journalists in one region—the western-most northern part of Canada—reported on other northern regions, and how others covered Yukon news (in short, *rarely*).

From 1990 to 1996, I conducted a case study of newspaper representaions of Indigenous people. I altered the conventional, quantitative approach to content analysis to bring in qualitative elements, using numerical analysis to supplement qualitative observations because the numbers in such a study cannot stand on their own. This differs from the usual reporting of data as "objective" and solitary figures. I included reassessments of the same data by different researchers and extensive discussions of the researchers' perspectives, biases, and assumptions, and of the limitations inherent in collecting and analyzing data of this kind (Alia 1996). A member of the research team, Brian Higgins, analyzed the data for a chapter of *Un/Covering the North* (Higgins and Alia 1999). It was a time of forward leaps and backward maneuvers, and contradictory funding practices that encouraged quantum improvements to broadcasting and discouraged similar shifts in print. One of the three newspapers in the study, the Northwest Territories First Nations paper, *The Press Independent*, succumbed soon after the study was completed. The Nunavut weekly, *Nunatsiaq News*, was able to report at close hand on the process leading up to the 1999 launch of Canada's newest territory, Nunavut. Today, *Nunatsiaq News* reports from Nunavut, to an international online audience, as well as continuing its life as a print weekly.

From 1990 to 1995, I interviewed more than 100 people who agreed to speak on the record, and about 30 others who were guaranteed anonymity. Occasionally, I became an unintentional participant observer. I had planned a day observing debate in the Northwest Territories Council chambers in Yellowknife. Shortly after the start of debate, someone dashed into the chamber with information about a mine explosion. Everything stopped, while people dealt with the crisis. I became an impromptu, volunteer reporter for *Native Press*—I had journalism experience; the newspaper staff was small; they needed help. The mine disaster eclipsed other stories and exaggerated outside media attention. The "skewed" coverage revealed a pattern I have observed from 1985 to the present: every crisis involving Indigenous people brings them temporarily closer to outsiders. The study's time frame was set and the research long begun, before the explosion at Giant Mine that caused the deaths of nine men. While it could be argued that another time period would reveal more "typical" coverage, I think the study does reveal typical coverage. Northern and Indigenous people receive coverage in "mainstream" media *primarily* in times of crisis, a point borne out by the study, which encompasses coverage before, during, and after the explosion.

The research process also included two years as a participant observer and volunteer at the Inuit Tapirisat of Canada (now Inuit Tapiriit Kanatami) head office in Ottawa. I was privileged to engage in continuing discussions with ITC staff and to have access to the material archived there. In return, I did research, writing and editing on various reports and policy documents, and helped start a newsletter, *Kakivak*. Throughout the period from 1982 to 2009, formal interviews were supplemented by consultation with Indigenous and non-Indigenous media practitioners, academics, political and community leaders, media consumers, and others, privately and at workshops and conferences in several countries and regions.

Abbreviations

AAAS	American Association for the Advancement of Science
ABC	Australian Broadcasting Corporation
ACECSOGUA	Associación Coordinadora de Emisoras y Comunicadores de Sur Occidente de Guatemala
AINA	Arctic Institute of North America
AIROS	American Indian Radio on Satellite distribution system (see NAPT)
AMARC	World Association of Community Radio Broadcasters
AMECOS	La Associación de Medios Comunitarios y Comunicadores Sociales
AMMSA	Aboriginal Multi-Media Society (Canada)
ANCSA	Alaska Native Claims Settlement Act (USA)
APAG	Amerindian Peoples Association of Guyana
APFT	Avenir des Forêts Tropicales
APTN	Aboriginal Peoples Television Network (Canada)
ARCG	Associación de Radios Comunitarias de Guatemala
ATSIC	Indigenous Broadcasting Policy Australia
ATV	All-terrain vehicle (three-wheel vehicle used in the Canadian north)
AVR	Aboriginal Voices Radio (Canada)
BBM	Bureau of Broadcast Management
BRACS	Broadcasting Remote Aboriginal Community Scheme (Australia)
CAAMA	Central Australian Aboriginal Media Association
CAC	Caribbean Amerindian Centrelink
CAEPR	Center for Aboriginal Economic Policy Research

CAMCO	Canadian Aboriginal Media Co-Operative
CBC	Canadian Broadcasting Corporation
CDI	Comision Nacional para el Desarrollo de los Pueblos Indígenas de México
CEFREC	Centro de Formación y Realización Cinematográfica (Bolivia)
CGCC	Consejo Guatemalteco de Comunicación Comunitaria (Guatemala)
CIDOB	Bolivian Indigenous Peoples Audiovisual Council
CIER	Centre for Indigenous Environmental Rights
CMCs	Community multimedia centers
COCOC	United Nations Economic and Social Council
CONAIE	La Confederación de Nacionalidades Indígenas del Ecuador
CONFENIAE	Confederación de Nacionalidades Indígenas de la Amazonia Ecuatoriana
CRTC	Canadian Radio-television and Telecommunications Commission
CSCB	Bolivian Settlers syndicate Confederation
CSUTCB	Bolivian Rural Workers Sole Syndicate
CVI	Centro de Video Indigena (Mexico)
CYC	Company of Young Canadians
DAB	Digital audio broadcasting
DR	Danmarks Radio (Denmark)
DTH	Direct-to-home satellite service
DTN	Delay Tolerant Network
DVC	Digital Video Compression
ECUARUNARI	Confederación de Pueblos de la Nacionalidad Kichua del Ecuador
FCUNAE	Federación de Comunas Unión de Nativos de la Amazonía Equatoriana
FENOCIN	Federación Nacional de Organizaciones Campesinas, Indígenas y Negras
FICI	Federación Indígenas y Campesino de Imbabura
FNEN	First Nations Environment Network (Canada)
FNVC	Fiji National Video Centre
FOCIFC	Federación de Organizaciones Indígenas de la Faldas del Chimborazo
FPP	Forest Peoples Program

GFIPIS	Global Forum of Indigenous Peoples and the Information Society
HBC	Hudson's Bay Company (Canada)
IAITPTF	International Alliance of Indigenous and Tribal Peoples of the Tropical Forests
IBC	Inuit Broadcasting Corporation (Canada)
IBIN	Indigenous Peoples Biodiversity Information Network
ICC	Inuit Circumpolar Council (formerly Inuit Circumpolar Conference)
ICCI	Instituto Cientifico de Culturas Indígenas
ICCTA	Indigenous commission for Development of Communications Technologies in the Americas
ICT	Information and Communications Technology
ILO	International Labor Organization
INN	International Native News
INnet	Indigenous Issues Network
IPBN	Indigenous Peoples Biodiversity Network
IPDC	International Programmed for the Development of Communication
IPS	Inter Press Service news agency
ITC	Inuit Tapirisat of Canada (now ITK, see below)
ITK	Inuit Tapiriit Kanatami
ITU	International Telecommunications Union
ITV	Indigenous Television Network (Taiwan)
IWN	Indigenous Women's Network
JAWS	The Journalist Association of Western Samoa
KNR	Kalaallit Nunaata Radioa (Greenland)
LAMP	Living Archive Management Project (South Africa)
LANIC	Latin American Network Information Center
LGA	Local Government Act (USA)
MICC	Movimiento Indígena y Campesino de Cotopaxi
MMSEA	Ethnic Minorities and Networks in Mainland Montane, Southeast Asia
MRG	Minority Rights Group International
NAIBA	National Aboriginal and Islander Broadcasting Association (Australia)
NAJA	Native American Journalists' Association (US and Canada)

NAN	Native American News
NAPT	Native American Public Telecommunications (US)
NARF	Native American Rights Fund
NCRF	National Community Radio Forum (South Africa)
NCS	Native Communications Society (Canada)
NFB	National Film Board of Canada
NGO	Non-governmental organization
NIC	Nunavut Implementation Commission
NIMAA	National Indigenous Media Association of Australia
NIPA	Native Indian/Inuit Photographers' Association (US-Canada)
NIRS	National Indigenous Radio Service (Australia)
NIT	National Indigenous Times (Australia)
NNBAP	Northern Native Broadcast Access Program (Canada)
NNBY	Northern Native Broadcasting Yukon (Canada)
NNN	Native Nations Network (Canada-USA)
NPA	North People Academy (Russia)
NRK	Norwegian national broadcaster
NWT	Northwest Territories
OMVIAC	Organización Mexicana de Videoastas Indígenas (Mexico)
OPIP	Organización de Pueblos Indígenas de Pastaza
OPM	Free Papua Movement
PACNEWS	Pacific News Agency
PacTOC	Pacific Telecentre Online Community
PEACESAT	Pan-Pacific Education and Communication Experiments by Satellite
PIBA	Pacific Islands Broadcasting Association
PINA	Pacific Islands News Association
PIPE	Partnership for Indigenous Peoples' Environment
PJNP	Programme in Journalism for Native People (Canada)
PNG	Papua New Guinea
RAIPON	Russian Association of Indigenous Peoples of the North
RATNET	Rural Alaska Television Network
RCAP	Royal Commission on Aboriginal Peoples (Canada)
RCERPF	Royal Commission on Electoral Reform and Party Financing (Canada)

RCMP	Royal Canadian Mounted Police
RCTS	Remote Commercial Television Service (Australia)
RIBS	Remote Indigenous Broadcasting Service (Australia)
ROM	Royal Ontario Museum (Canada)
RSF	Reporters sans Frontières
SAIIC	South and Meso American Indian Rights Center
SNC	Sámi Network Connectivity Project
SRCI	Sistema de Radiodifusionas Culturales Indigenista (Mexico)
TBC	Tongan Broadcasting Commission
TEABBA	Top End Aboriginal Bush Broadcasting Association (Australia)
TOW	Telecentre on Wheels (India)
TVNC	Television Northern Canada
TVNZ	Television New Zealand
UCTP	United Confederaton of Taino People
UNESCO	United Nations Educational, Scientific and Cultural Organization
UNORCAC	Unión de Organizaciones Campesinas Indígenas de Cotacachi
UNPO	Unrepresented Nations and Peoples Organization
UPAI	Indigenous Audiovisual Production Units (Mexico)
USET	United South and Eastern Tribes, Inc. (USA)
VCD	Video compact disc
WAPHA	Women's Alliance for Peace and Human Rights in Afghanistan
WBREDA	West Bengal Renewable Energy Development Agency
WCIP	World Council of Indigenous Peoples
WIMSA	Working Group of Indigenous Minorities in Southern Africa
WITB	World Indigenous Television Broadcasting
WSIS	World Summit on the Information Society
WWF	World Wildlife Fund
YLE	Yleisradio (Finland)

Introduction

How I Came to Be Here

This is not my story. My ancestry is East European-North American and not, to my knowledge, Indigenous. However, ethnicity is seldom sufficient to explain one's interests and convictions. Let me briefly clarify how I came to be here.

I was born in New York and raised in Oklahoma in the 1950s. It was a time of rapid change—from oppression and discrimination to sociopolitical unrest, activism, and change. Oklahomans annually celebrate the advent of statehood on "'89ers Day." When I was a child, this meant "White Christian People's Day," and dressing up as "pioneers." African American, Hispanic, Jewish, and other ethnic- and religious-minority Oklahomans tended to feel less celebratory about their arrival as slaves, migrant laborers, refugees or migrants who were variously abused and unwelcome. For Native Americans, Oklahoma was the final stop at the end of what has become known as the Trail of Tears—the forced migration and consequent decimation of an array of Indigenous peoples that was aimed at clearing them from more desirable lands. I grew up alongside the myths of "American settlers" of "the West." It took a long time to learn that "settler" often meant "stealer of other people's lands."

As a child, I was regularly attacked by other children, who sometimes called out, "You killed our Lord," while beating me. Being a transplanted "Yankee" was considered almost as bad as being Jewish. On the playground, the US Civil War was re-enacted daily, along with "Cowboys and Indians" play that set Native Americans in continuing opposition to "real" Americans—all of the children being descendants of relative newcomers to the North American continent. Oklahoma practiced what in South Africa was called *apartheid*. Public buses had prominent signs saying "Colored please step to the rear." Public places had separate toilets and drinking fountains labeled "Colored" and "White" and separate (and unequal) schools. My family was harassed for being Jewish, for advocating equality, and for including people who then called themselves Negroes among those who visited our home. There were threatening phone calls; hate literature regularly appeared at my father's downtown office, where he served as executive director of the Jewish Community Council.

In the peculiar hierarchy of 1950s Oklahoma, Native Americans were less often derided than exoticized. I had Native American and ethnic minority classmates, but African Americans attended only "colored" schools. Recently, while cleaning out old papers, I came across an '89ers Day program from the University Elementary School, run by the University of Oklahoma in Norman. It was more progressive than the public schools (I was one of the first girls to take wood shop, in what was considered a radical experiment). The '89ers Day celebration featured "Cowboys" and "Indians"; the silk-screened, black and brown, program cover includes both. Even so, the "cowboy" is on the front and the "Indian" is on the back and both (of course!) are men. Both men have brown faces—though incongruously, the "cowboy's" hands are white, while the "Indian" has brown hands and chest—and what looks for all the world like a "caveman" style, one-shoulder, leopard-skin robe. Both are stereotypes, but the "Indian" is more "primitively" dressed.

It was 22 April 1954 and I was twelve years old. We sang "Songs of the '89ers"—"Prairie Schooner," "Wagon Wheels," and "Home on the Range." The show's three acts were called "NEW LAND AND HOMES," "GOOD TIMES ON THE PRAIRIE," and "BRAND NEW STATE." There was no indication of whose land and homes we were singing about. Among the 14 numbers was a song called "Shoot the Buffalo" and another called "Buffalo Head Dance," of unknown (but presumably Native American) cultural origin. The children were divided into two groups of "Indians" and "Cowboys," set into "INDIAN TERRITORY" and "OPEN RANGE." Most children were in the chorus. None of the "Cowboys" were girls. In fact, only two girls appear on the program: Lucia Rohrbaugh and Valerie Graber (me!). We are both in the "INDIAN TERRITORY" section, and—to my eternal horror—are designated "Squaws." For the record, the University of Oklahoma is now one of the outstanding publishers of books by and about Native Americans.

If that was 1950s Oklahoma's more enlightened side, you can imagine the rest. Not long after returning from Norman to school in Oklahoma City, I became involved in the Civil Rights Movement. I served on the National Conference of Christians and Jews youth steering committee, spoke in public and on radio and television. I marched, met, strategized, supported, and participated in lunch counter sit-ins, including the pivotal ones at Katz's Drug Store and Brown's Department Store in Oklahoma City. Much of my learning took place out of school; I was taught by an amazing group of powerful women: Clara Luper (teacher, activist, and civil rights leader), Hannah D. Atkins (the first African American woman elected to the Oklahoma House of Representatives, later named by US President Jimmy Carter to the UN General Assembly), Evelyn La Rue Pittman (composer, choir director, and champion of African American music) in Oklahoma; and Jenny Wells Vincent, concert pianist-turned folk singer who pioneered in using songs to teach language and culture across New Mexico's Spanish American, Native American, and Anglo communities. Among numerous, mostly unsung accomplishments, my father quietly broke the "color barrier" by organizing a major social work convention at Oklahoma City's main hotel. When the owner said, "OK," but "Colored will have to use the rear entrance," he replied that African American participants would use the front door with the other delegates, or the convention could take its business elsewhere.

Music was always important in my life. My earliest memories are of listening to 1978 recordings of Paul Robeson. Between New York and Oklahoma, we spent a couple of years in Phoenix, Arizona, where, at age four, I had a formative experience. Robeson was scheduled to sing; my father, Julius Graber, was his designated host. We attended the concert. As he often did, Robeson gave a free, second concert at a union hall, after which my dad brought him back to our home to relax. As if I had always known him (which in a way, I had), I climbed onto his lap and he sang me the Russian lullaby I had heard and loved since birth. I can still recall the layout of our living room, the chair, his wonderful smile, the enormity of the man ... that Voice. His life and work continue to inspire; I remain angry and saddened by the shameful treatment he received at the hands of official America.

In the 1980s, I worked with colleagues at York University and several Ontario First Nations on a communications networking project. In the early '90s, Bud Whiteye and I developed the First Nations Intensive seminar for University of Western Ontario journalism students. Bud founded Native News Network of Canada, on whose board and advisory board I proudly served. When I met Pete Steffens in 1996, things came full circle. Pete had worked for decades with African American, Hispanic, and Native American people, particularly members of the Lummi Nation in Washington State. His own story features lifelong political and social consciousness. His parents were the great "muckraker," Lincoln Steffens, and the activist-writer, Ella Winter. Pete's dad was one of my own dad's heroes; Lincoln Steffens was often quoted in our home.

His works are enjoying a revival; it is perhaps unfortunate that they are still so timely.

Having reached the maturity that is supposed to bring understanding, I remain perplexed by the persistence of inequality and discrimination, and stunned by the achievements of people who struggle against what often seem insurmountable odds. It is with the greatest respect and admiration that I have watched the emergence of The New Media Nation.

In 2008, while chatting over coffee, I told an otherwise well-educated woman that we had just attended the North American Indigenous Games in Duncan, British Columbia. When I mentioned the Indigenous media booths, she sounded surprised and said, "What a shame, there's almost no Indigenous media." I have heard similar comments in Europe, the United Kingdom, and the United States. The thousands of people worldwide, who are creating and energetically using Indigenous media, know how mistaken she was. I hope this book will help to spread the word.

Indigenous-run media do not just fill the airwaves, they carry new worldviews, in sometimes unexpected ways. One recent afternoon while watching the weather report on APTN, the Aboriginal Peoples Television Network, I realized that a new map of Canada has been created, quietly and without fanfare. The station shows weather conditions and forecasts for small, remote communities that are almost never included on other channels. On APTN's maps and reports, communities of importance to Indigenous people receive the same prominence as major cities; it is people, not population size that counts. Thus, the viewer understands the tiny Nunavik community of Inukjuaq to be as important as the urban centers of Ottawa and Montreal. Among the communities having a substantial presence are those of Yukon, a fitting tribute to a small territory that has made a disproportionately large contribution to Indigenous politics and media. As of December 2007, the territory's population totaled 32,714 (Government of the Yukon 2008). Northern Native Broadcasting Yukon (NNBY) was one of the organizations that founded Television Northern Canada and later, the Aboriginal Peoples Television Network. The illustration that follows shows a team from NNBY's television service on location.

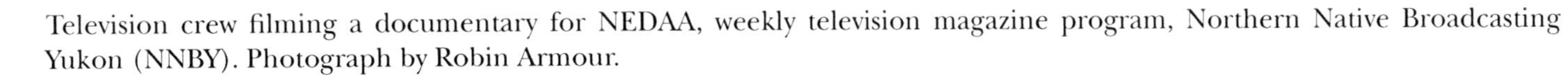

Television crew filming a documentary for NEDAA, weekly television magazine program, Northern Native Broadcasting Yukon (NNBY). Photograph by Robin Armour.

CHAPTER 1

Scattered Voices, Global Vision

We signed the papers with our thumbs.
As soon as the ink was dried, the surrounding
peoples challenged our treaties
and the fight began.
They told us to be self-sufficient after secluding us
on a rock pile
—Robert Joe, Sr. [Wa-Wal-Ton], Swinomish [nation]

Some of the world's least powerful people are leading the way toward creative and ethical global media citizenship. Locally, regionally, nationally, and internationally, Indigenous peoples are using radio, television, print, and a range of new media to amplify their voices, extend the range of reception, and expand their collective power. Emerging from the shadows of a shared colonial inheritance, the international movement of Indigenous peoples has fostered important social, political, and technological innovations.

I first used the term, "New Media Nation," in a chapter for Karim Karim's edited collection on communication and Diaspora (Alia 2003). Emerging from the international movement of Indigenous peoples, The New Media Nation is linked to the explosion of Indigenous news media, information technology, film, music, and other artistic and cultural developments. While its individual member outlets and organizations may be subject to state regulations and control, in a broader sense, The New Media Nation is an outlaw organization. No real "nation" in the political science sense, it exists outside the control of any particular nation state, and enables its creators and users to network and engage in transcultural and transnational lobbying, and access information that might otherwise be inaccessible within state borders.

Like many of its constituent organizations and cultural groups, The New Media Nation uses a form of what the interdisciplinary, post-colonial scholar of ethics, human rights, and globalization, Gayatri Chakravorzty Spivak (1990; 1995), has called "strategic essentialism," in which particularities and differences are set aside, in the interest of constructing an essentialized pan-indigeneity. Cultural action, the making and remaking of identities, takes place "in the contact zones, along the policed and transgressive intercultural frontiers of nations, peoples, and locales (Clifford 1997: 7)." Transnational connections break the binary relations of "minority" and "majority", and renew earlier concepts such as W. E. B. Du Bois's notion of "double consciousness" (Clifford 1997: 255). Those who exoticize and primitivize Indigenous peoples may find these developments surprising. The Inuit journalist, Rachel Qitsualik, provides a cultural context for understanding the openness of Inuit and other Indigenous communicators to technological innovation and change:

> Inuit are nomads…[and] rejoice in the ability to compare Opinions abroad, as they did when traveling at will…the hamlet is the new iglu, and the Internet is the new Land (Alia 2007).

Through creative use of strategic essentialism (Spivak 1995) and a wide range of media techniques and technologies, Indigenous people are developing their own news outlets and networks, simultaneously maintaining or restoring particular languages and cultures and promoting common interests. Their progress is consistent with Ien Ang's idea of the "progressive transnationalization of media audiencehood" (Ang 1996: 81). However, "transnationalization" implies a unidirectional crossing of national boundaries and, in my view, should be extended to account for instances of internal colonialism and for boundaries between ethnicities and regions. I have called the fluid, constantly changing crossing from boundary to boundary and place to place—the *inter*nationalization of Indigenous media audiencehood and media production—"The New Media Nation."

Inuit have played an important role in this global communications movement, having developed some of the world's most effective and politically astute organizations. Indigenous people from many parts of the world regularly attend Inuit Circumpolar Council (ICC) assemblies as observers who often carry ideas and strategies home. Founded in 1977 by the late Eben Hopson of Barrow, Alaska, as the Inuit Circumpolar Conference, the ICC has flourished and grown into a major international non-government organization representing about 150,000 Inuit of Alaska, Canada, Greenland, and Chukotka (Russia). The organization holds Consultative Status II at the United Nations. On its website, the ICC explains its origin and mandate:

> To thrive in their circumpolar homeland, Inuit had the vision to realize they must speak with a united voice on issues of common concern and combine their energies and talents towards protecting and promoting their way of life.
>
> The ICC international office is housed with the Chair and each member country maintains a national office under the political guidance of a president (ICC website 2006).

The ICC declared its global vision from the start. During the Cold War years, Siberian Inuit were not able to leave the Soviet Union to attend ICC assemblies. But ICC was founded "under four flags" and each assembly flew them all. Each time, an empty chair was placed in front of the Soviet flag to symbolize the presence of Inuit from all of the circumpolar nations. Finally, in 1989, under Gorbachev's leadership, Inuit from Chukotka, Siberia, were permitted to attend the ICC assembly. It was held in Sisimiut, Greenland, and I was lucky enough to be there for one of the most moving experiences I have known.

Everyone gathered on the dock to greet the ferry when the Siberians arrived. Though happy, some of them were a bit wobbly, having reportedly endured 40-foot waves during the passage from the airport at Sondre Strømfjord to Sisimiut. Some of their hosts went out to greet the ferry in the large boats called *umiaqs.* The magnificent Greenland *qayaq* (kayak) team performed in their honor. Ashore, they were met with tears and laughter. They were granted provisional status and brought into ICC as full members at the next assembly. Every evening there was dancing, theater, and music. Inuit from Alaska and Siberia watched each other's performances and expressed their joy in meeting relatives for the first time. Some had been separated by only a short distance, through the boundaries set by governments that had long denied their cultural and often familial connection in the name of political expediency. In 1989, there were several chairs in front of the Soviet flag, and all of them were full. These days, the flag is Russian.

Figure 1.1. Mary Simon addressing the ICC General Assembly in Sisimiut, Greenland, 1989. To her left are delegates from Chukotka, Siberia (USSR, now Russia), who were attending the assembly for the first time. Photograph by Valerie Alia.

Rewriting the story

> Recently, the spread of information and communication technology has given rise to new blends of traditions and elements of "world culture," in music and arts, clothing fashions, food, etc. The coexistence of traditions and modernity is currently observed among many indigenous peoples worldwide. It includes renewal and in some cases reinforcement of ethnic identities, as well as an instrumentalization and commoditization of cultures...These new blends illustrate Marshall Sahlins's assertions, based on research in many parts of the Third and Fourth Worlds, that peoples around the world see no opposition between tradition and change, indigenous culture and modernity, townsmen and tribesmen. Culture is *not* disappearing, he concludes, rather it is modernity that becomes indigenized.
> —Csonka and Schweitzer 2004: 51; Sahlins 1999

One example of Sahlin's view of indigenized modernity is a delightful booklet of historical and cultural information, which was published in 1989 by the Dene Cultural Centre in Yellowknife, Northwest Territories (Canada). The reader encounters indigenized modernity from the moment of seeing its front cover, with a picture of Indigenous people on a riverbank, watching a canoe go by, filled with strangers. One child is pointing to the canoe. The title fills in the rest of the picture—*Dehcho: Mom, We've Been Discovered.* Steve Kakfwi is a political leader from the Dene Nation. The back cover of the booklet carries his words:

> Alexander Mackenzie came to our land. He described us in his Journal as a 'meagre, ill-made people'...My people probably wondered at this strange, pale man in his ridiculous clothes, asking about some great waters he was searching for. He recorded his views on the people, but we'll never know exactly how my people saw him. I know they'd never understand why their river is named after such an insignificant fellow (Kakfwi, quoted in Holmes 1989).

The Dene call the river Dehcho, not the Mackenzie River. The booklet's text, design, and presentation make Sahlins' point. Colonial and Indigenous history and culture are rewritten and presented for the education (and perhaps, the edification) of Outsiders and the gratification of Insiders in setting the record straight. The humor, too, is characteristic.

It appears in all sorts of contexts and places, and offers a fresh way of communicating the issues. Another notable example is the Indigenous Australian film, *Babacueria* (1981), in which colonizers and the colonized are also turned around. The title is a satirical reworking of outsiders' hearing of the term, "barbecue area". In the film, a boatload of Indigenous "explorers" land on Australian shores and "discover" pale-skinned, English-speaking "natives." The explorer-colonizers proceed to do all of the things for which colonizers are known. They plant a flag, attempt to communicate with the "natives," rename and label people, things, and places in absurd and inappropriate ways, and then begin to "improve" the people they have found. From broadcast news anchors to government leaders and social service directors, they set new policies and controls, herd children off to residential schools, and engage in all manner of amusing and sometimes sinister activities.

The Fourth World Movement

The New Media Nation is one of the major trends to emerge from what George Manuel called the "Fourth World." While the term is generally attributed to Manuel and is unquestionably linked to his vision of an international organization of Indigenous peoples, Mbuto Milando apparently coined it during a conversation. When they met, Milando was First Secretary of the Tanzanian High Commission. Manuel told him that if Indigenous peoples could stand together, their voices "could reach beyond their national borders and be heard by the world. Mbuto...suggested that what George Manuel seemed to be talking about was the emergence of a "Fourth World." George Manuel seized on the phrase and over the years popularized it" (McFarlane 1993: 160–161).

George Manuel came from the Shuswap (Secwepemc), Neskonlith First Nation of British Columbia, Canada. He was equally passionate about working locally, regionally, nationally, and globally. Regionally, he served as president of the Union of British Columbia Indian Chiefs. He headed the National Indian Brotherhood (now the Assembly of First Nations). He was the first president of the World Council of Indigenous Peoples. In the late 1970s, he told 15,000 Indigenous Peruvians, "we have to have our own ideology" (Manuel and Posluns 1974: 77–78). His influences were as wide as his outlook. His thinking was affected by meetings with Julius Nyerere in Tanzania and with Māori leaders in New Zealand in the early 1970s. In 1973, he met in Washington, DC, with Mel Tomasket, president of the National Congress of American Indians. They set up exchanges between Indigenous people in Canada and the United States that became models for later developments in global and pan-Indigenous projects. Manuel was "'instrumental in drafting the Universal Declaration of the Rights of Indigenous Peoples' and contributed to innumerable and highly influential national and international position papers and reports. To help unify Indigenous peoples of North, Central, and South America and Eurasia, he launched the UN-affiliated World Council of Indigenous Peoples in 1975" (McFarlane 1993: 8).

It was not his first international effort. In 1960, he and Henry Castilliou submitted a brief on British Columbia land rights to the United Nations. In 1973, he traveled with a delegation from Canada's Department of Indian Affairs to New Zealand and Australia. He found a historic colonial link between First Nations people of British Columbia and Indigenous peoples in Eurasia: in both regions, the first European to "make contact" was Captain James Cook. Both areas were subjected to settlement and "developed under the British imperial system" (McFarlane 1993: 156). Manuel's tour of New Zealand ended with festivities organized by Māori that included members of some of the Polynesian peoples who had immigrated to New Zealand—Samoans, Tahitians, and Fijians, who shared a common culture with Māori (McFarlane 1993: 159). Before the time of PowerPoint, George Manuel and Michael Posluns created a narrated slide show of Manuel's Australia-New Zealand trip to promote the North American-Eurasian Indigenous project. In October 1975, the Nootka community hosted the first World Council of Indigenous Peoples (WCIP) conference on Vancouver Island. In attendance were 52 delegates and more than 200 official observers from 19 countries. George Manuel was acclaimed President. Vice President was Sam Deloria, from the United States. The rest of the founding executive comprised Julio Dixon (Panama), Clemente Alcon (Bolivia), Nils Sara (Norway), and Neil Watene (New Zealand) (McFarlane 1993: 218).

Manuel used the term "Fourth World" to clarify the position of Indigenous Peoples in relation to the layers of dominance and subordination, and centrality and marginalization of peoples within political power structures in the relatively privileged "First" and "Second" worlds and the "Developing" or "Third" world (Manuel and Posluns 1974; Bodley 1999: 77–85; Dyck 1985). Peoples of the "Fourth World" cannot be confined within national or state borders. For Manuel, language and communication help to engineer and maintain the oppression of Indigenous peoples. In the context of Fourth World theory, "Nation" refers to a community or group united by common descent and/or language (Murphy 2000). Manuel framed the Fourth World not as a place, but as a global highway. "The Fourth World is not…a destination. It is the right to travel freely, not only on our road but in our own vehicles" (Manuel and Posluns 1974: 217).

Fourth World: The Next Generation

However powerful a movement, progress tends to come in fits and starts. In 2007, "the new UN Human Rights Council adopted a declaration... to protect the rights of indigenous peoples around the world, including their claims on land and resources" (Schein 2007). While the declaration is not legally binding, many people have noted that the Universal Declaration of Human Rights, which also was not a binding document, eventually became customary law.

Ellen Lutz, executive Director of Cultural Survival, thinks the question of Indigenous "peoples" was "the main stumbling block... As indigenous advocates frequently point out, the whole debate is over the letter 's'" (Lutz 2007: 14). I think it is a bit disingenuous to imply that "Indigenous peoples" form as easily definable a category as "women" or "victims of disappearance." In fact, the very question of whether or not to define indigeneity underlies the whole enterprise. There is no definition of "Indigenous peoples" in the Declaration, or in any other UN publication, body or procedure. Lutz says that this is deliberate, as a precise definition is both impossible and unhelpful. "Among indigenous peoples, inter-governmental organizations, and most states, consensus has emerged that it is better not to define 'indigenous peoples' and to focus instead on defining and protecting their rights" (Lutz 2007: 19).

Despite the frustration of knowing the decision was not unanimous, participants acknowledged it as an emotional moment of great political significance. "[T]he packed conference room erupted into applause. People wept and hugged each other and smiled broadly. Louise Arbour, the UN high commissioner for human rights and former Supreme Court of Canada justice, joined in the standing ovation" (Schein 2007). "In the end, the force for support was so great that even countries that had previously dissented, including Russia and Colombia, chose to abstain rather than vote against it. The only 'no' votes were from the diehard group of Canada, Australia, New Zealand, and the United States" (Cultural Survival Quarterly 2007: 5). In April 2009, under the leadership of newly elected Kevin Rudd, the Australian government reversed its position and endorsed the declaration. Despite a sharp change in direction from the presidency of George W. Bush to that of Barack Obama, the United States still had not endorsed the declaration (Cultural Survival 2009). For many, Canada's vote was a big shock. The newly (and only marginally) elected minority Conservative government of Canada rejected the request for a consensus decision, requested a vote, and then voted against the declaration.

Lawyer and Treaty Six international chief, Willie Littlechild said, "I'm very excited...delighted and encouraged by the signal the new Human Rights Council has given the world that they are serious about addressing indigenous issues." Nevertheless, Littlechild and Kenneth Deer, representing the Mohawk Kahnawake First Nation of Québec and the United Nations Council of Chiefs, said they felt betrayed by the Canadian government. Deer said: "Canada had a lot to do with the declaration getting this far ... It's ironic that for eleven years they carried the resolution and at the end they voted against the declaration and against their own work" (Schein 2007).

The New Media Nation is a Fourth World movement that is engaged with removing ethnic and national borders and placing pan-indigeneity at the center. Although culturally distinct, the world's Indigenous communities have collectively experienced many of the elements of Diaspora. Small numbers of people are scattered over great distances, some far from their homelands, as in Oklahoma—where survivors of forced

Figure 1.2. The next generation: high school students join the staff of the Top End Aboriginal Bush Broadcasting Association (TEABBA), broadcasting live from the Gama Festival in Arnhem Land, Australia in 2005. Photograph by Michael Meadows.

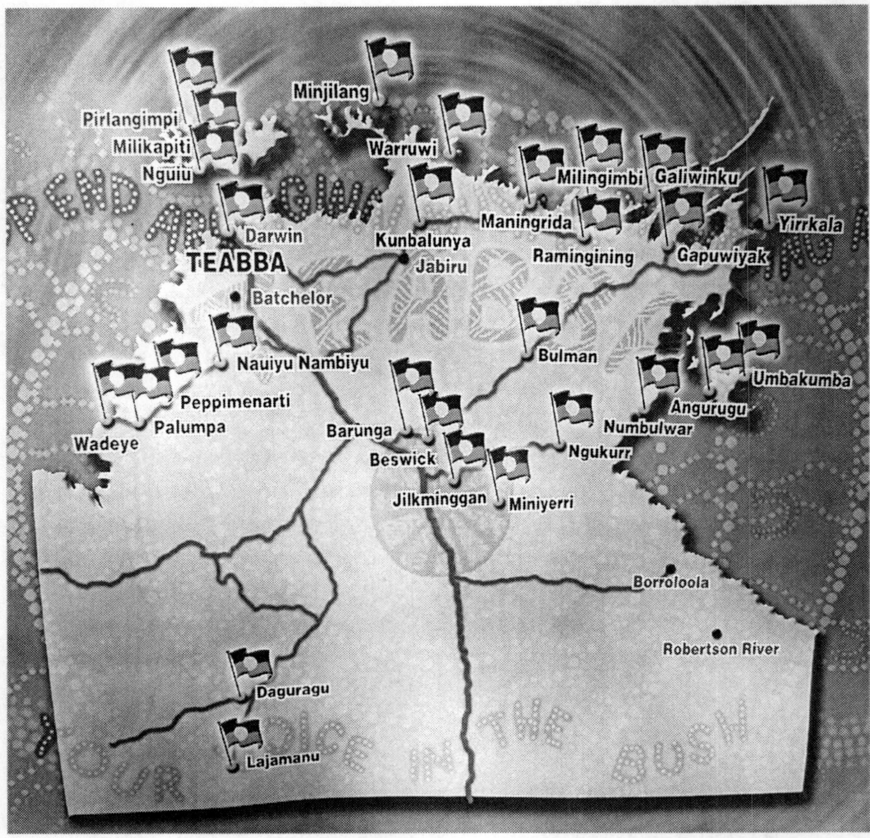

Figure 1.3. Radio stations served by TEABBA, across the Top End of Australia.

relocation landed at the end of the "Trail of Tears," and the high Canadian Arctic, where Inuit were moved from northern Québec. Some reside in homelands newly "legitimated" by dominant governments—as in the instances of Nunavut Territory and Greenland Home Rule. Some are in process of negotiating or renegotiating their relationships to state governments, as in the case of Indigenous people in Australia, in the wake of the 2007 government apology (discussed in Chapter 2). Arjun Appadurai (1990: 298–299) uses the term, mediascape, to describe media representations and their dissemination. Starting from Appadurai's idea of "mediascapes" and "ethnoscapes"—new flow patterns of media and people—John Sinclair and Stuart Cunningham observe that "whereas flows of people often have tended to be from…the periphery…towards the "centre", media flows historically have traveled in the other direction" (Sinclair and Cunningham 2000: 2). This is not always the case with respect to Indigenous media. Although dominant-society media made early incursions into Indigenous communities, the main movement in Canada has been from "periphery" to "core"—with Indigenous media originating in remote arctic and sub-arctic communities and moving gradually toward the urban centers. The most recent example of this pattern is the transformation of Television Northern Canada (TVNC), based in the minimally populated northern regions, to the Aboriginal Peoples Television Network (APTN), which covers all of Canada and is moving toward global coverage. The poverty and distortion of mainstream coverage have made it imperative for Indigenous people to develop their own news outlets. They are using satellite, digital, cable, the Internet, cell phones—whatever is at hand. This is not, as some might infer, an exclusively high-tech movement, but one based on maximizing the effectiveness of small- and large-scale technologies, organizations, communities, and budgets. Often, it is the older and simpler technologies—or a fluid mix of old and new, high- and low-tech—that best serve its needs.

Radio remains the chosen medium for local communication, both in traditionally transmitted forms and transmitting via the Internet to expand and globalize originally localized broadcasts. As access increases and technologies evolve and blend, the Internet is fast becoming the second medium of choice, with websites, blogs, social networking, chat rooms, mobile phones, and radio intertwined within it. The Internet is the primary outlet for broader interactive communication, a forum for discussion and debate, and a tool for global and regional constituency building and cross-border organization. Nevertheless, radio persists, despite the increased availability of digital television, video-on-demand, and iPod. Its audiences are growing, not receding, alongside the growth of other media. In August 2007, *The Guardian* reported that:

> the latest audience figures published yesterday reveal that we are more in love with the radio than ever before…[but] we are not listening to it in quite the same way as we used to…radio is now piggy backing on the digital revolution, with nearly 12 million people—26% of the adult population—tuning in via digital radio, digital TV and the internet.
>
> About 4.4 million listen on their mobile phone, up more than 25% on last year (Plunkett 2007: 3).

Youth often lead their technophobic elders. In 1996, a group of Sámi youth organizations launched a website providing daily updates of the Fourth World Indigenous Youth Conference. Other projects followed, including websites by the magazines *Samefolket* and *Min Aigi*, that circulate Sámi news and information across Sápmi and Siberia—examples of youth promoting their elders' publications, transmitting multigenerational communications to old and new audiences.

According to Jérome Bindé, cultural diversity "is in danger" and requires "balanced policies" that help to preserve linguistic diversity while also promoting "widely spoken languages." Indigenous languages are "the main medium of expression of aspirations, intimate desires, feelings and local life. They are indeed the living repositories of cultures" (Bindé 2005: 147, 152). Radio and the Internet provide the most consistent support for local and global efforts to retain, restore, and strengthen Indigenous languages. Such efforts are aided by Indigenous language type fonts developed by Apple and Microsoft—in North Sámi, Inuktitut (the main language of Canadian Inuit), Dogrib, Chipewyan, North and South Slavey—used mainly in Canada's western Arctic, where the Northwest Territories Community Access Program (CAP) teaches Internet skills and helps communities to develop their own websites.

Not everyone concerned about language loss recognizes the importance of media for developing and strengthening language programs. *Cultural Survival* is an Indigenous rights advocacy organization founded by David and Pia Maybury-Lewis. Its publication, *Cultural Survival Quarterly,* usually features informative, passionate, politically and culturally astute articles. The Summer 2007 issue, titled "The Last Word," is a call to action to help "bring the last of the first languages back to life" (Cultural Survival 2007: front cover). Executive Director, Ellen L. Lutz, makes an impassioned plea for Cultural Survival supporters to become language advocates (Lutz 2007: 3). While she lists a number of approaches, projects, and programs, there is no mention of the potential value of media for "Revitalizing critically endangered Native American languages" (Lutz 2007: 3). In fact, only one of several articles on language revitalization even mentions the media.

Alice Anderton, executive director of Oklahoma's Intertribal Wordpath Society, notes that "The Cherokee Nation offers online classes, and the Yuchis are developing computerized activities and CDs" and the Kiwat Hasinay Foundation dedicated to Caddo language has "digitized and archived old song and language records; and produced publications such as a bilingual children's storybook and a Caddo language phrasebook." Anderton's organization itself produces Wordpath, a weekly 30-minute public access television program on Oklahoma Indian languages (Anderton 2007: 23). Mainly the issue focuses on classroom work—the most productive are programs based on the highly successful Māori "language nest" model. It is disappointing that none of the discussion includes the value of radio, television, and "new media" in the promotion, teaching, and retention of Indigenous languages. Perhaps that is because the United States has provided so little support for Indigenous media. In Canada, where Indigenous media have long received both financial and policy support, the role of media is widely acknowledged and promoted.

Networking the "Nation"

Drawing on her own experience as a scholar of Māori descent with international and transnational connections, Linda Tuhiwai Smith sees networking as a central and crucial activity.

> Networking is a process which indigenous peoples have used effectively to build relationships and disseminate knowledge and information…
>
> Networking by indigenous peoples is a form of resistance…In many states police surveillance of indigenous activists and their families is common practice. In some states, such as Guatemala, the disappearance of indigenous peoples has also been common practice. In these contexts networking is dangerous.
>
> Networking is a way of making contacts between marginalized communities…[by] definition…[excluded] from participation in the activities of the dominant non-indigenous society, which controls most forms of communication. Issues such as the Conventions on Biodiversity or GATT, for example, which have a direct impact on indigenous communities, are not addressed by mainstream media for an indigenous audience. Indigenous peoples would not know of such agreements and their impact on indigenous cultural knowledge if it were not for the power of networking (Tuhiwai Smith 1999: 157).

Following that principle, Māori took networking to new heights in 2008, when they played a leading role in the creation of the World Indigenous Television Broadcasters Network (WITBN), the first global Indigenous television network, and invited the Indigenous world to come and celebrate. In 2007, they announced the forthcoming gathering:

> AOTEAROA 08/WITBC
>
> WORLD INDIGENOUS TELEVISION IS HERE!
>
> THE FIRST EVER GATHERING OF INDIGENOUS TELEVISION LEADERS FROM THROUGHOUT THE WORLD WILL BE HELD IN AUCK-

> LAND, NEW ZEALAND FROM MARCH 26–28, 2008. THE WORLD INDIGENOUS TELEVISION BROADCASTING CONFERENCE—WITBC '08—IS AIMED AT UNIFYING, STRENGTHENING AND PROMOTING INDIGENOUS BROADCASTERS (WITBN 2007).

The conference theme was "Reclaiming the Future." The mood was buoyant. Participants heard that "Indigenous broadcasting is making great progress all over the world and there is renewed pride in indigenous issues, cultures and native languages" and that the conference welcomed "all leaders, producers and planners involved in indigenous television world-wide." The featured event was the launch of a World Indigenous Television Broadcasters Network, WITBN. The host was Māori Television, New Zealand's national Indigenous broadcaster, which expressed its hope that "the conference will become a permanent, prestigious international event hosted by a different broadcaster each time" (Māori Television 2007a). The second WITBN conference was set for March 2010 in Taipei, Taiwan, with Taiwan Indigenous TV (TITV)/Public Television Service (PTS, Taiwan) as hosts, with further conferences scheduled for 2012 in Wales, with Welsh broadcaster, S4C, hosting, and 2014 in Canada, hosted by APTN (Horan 18 June 2008).

WITBN held an interim council meeting in 2008, in Galway, Ireland, hosted by Irish language television channel, TG4. The interim council included broadcasters from Australia, Canada, Fiji, Ireland, Norway, Scotland, South Africa, Taiwan, and Wales. The first regular Council meeting took place in 2009, in Sápmi, in Karasjok, Norway, hosted by NRK Sámi Radio (WITBN 2009). Also in 2009, WITBN launched a Programme Exchange Scheme, an international news-sharing initiative, and an Indigenous current affairs series. Developed by TG4 in Ireland, the exchange scheme provides four programs a year to WITB Council members. The inaugural Council members are: APTN (Canada), BBC Alba (Scotland), NRK Sámi Radio (Norway), S4C (Wales), TG4 (Ireland), TITV/PTS (Taiwan), and Māori Television (New Zealand)." The news-sharing project provides Council members with weekly, 30-minute programs produced by Māori Television. Alongside WITBN, the 2008 Wairoa Māori Film Festival paid tribute to Māori film pioneers, Witarina Harris, Don Selwyn, and Barry Barclay (Wairoa Māori Film Festival 2008), and the 2009 festival screened films from Canada, the US, Australia, and Bolivia on the theme of "Kia Tau Te Rangimarie (Peace in our Time)"(Wairoa Māori Film Festival 2009).

The keynote speakers at the inaugural WITBN conference were:

Chief Judge Joe Williams, Māori Land Court, and Chairperson of the Waitangi Tribunal

Hon. Dr. Michael, Deputy Prime Minister of New Zealand

Hon. Parekura Horomia, Minister of Māori Affairs, New Zealand

Simon Molaudzi, Head of Education for the South African Broadcasting Corporation (SABC)

Saul J. Berman, Global Lead Strategy Partner and Global Strategy and Change Services Leader at IBM Global Business Services, USA

Sylvia H. Feng, President and Chief Executive, Public Television Service (PTS) of Taiwan

Margaret Mary Murray, Head of Gaelic BBC Scotland, "The Indigenous Voice of the BBC"

John Walter Jones, S4C, Wales

Jean LaRose, Chief Executive Officer, Aboriginal Peoples Television Network (APTN), Canada

Jim Mather, Chief Executive of Māori Television, Aotearoa-New Zealand.

Pól Ó Gallchóir, Chief Executive, TG4, Ireland.

Patricia Turner, Chief Executive Officer, National Indigenous Television (NITV), Australia

Shaun Brown, Managing Director of SBS, Australia

Tuwhakairiroa Williams, Chairman, Te Patahi Paoho—the Māori Television Service Electoral College, one of Māori Television's two stakeholders, Aotearoa-New Zealand

Garry Muriwai, Chairman of Māori Television

I was surprised to see how few of the keynote speakers were female. The disproportion was especially disappointing and surprising, given the prominence of women throughout the history of Indigenous broadcasting, worldwide and within the individual countries, cultures, and regions. It was also disappointing to see the top-down structure—most of the speakers and participants were chief, or other high level executives. It would seem appropriate for the structure and organization of such a landmark event to more accurately reflect the prominence of grassroots participation and collective action and organization in Indigenous media. It was also surprising that none of the important Sámi broadcasters was represented on the plenary program (Māori Television 2007b).

However progressive "progress" may be, such new "songs" too often include backward-sounding notes. Gendered and socially stratified thinking are not exclusive to colonial cultures. However elaborate the technology, it is always run by humans. Celebrating new developments need not mean unquestioning adoration, nor is progress served by avoiding challenge. Mass media carry all of it—the triumphs and celebrations, disasters and frustrations—across human and digital divides.

The processes of "[c]ultural homogenization and social diversification…have taken place simultaneously in the world during the last several decades…facilitated and complemented by the media" (Levo-Henriksson 2007: 1). The media are able to "diffuse group identities" by creating what Joshua Meyrowitz calls "'a *placeless* culture'…blurring the dividing line between private and public, and thus separating the traditional link between physical and social space" (Levo-Henriksson 2007: 2–3). Regardless of whether "its structural boundaries remain intact, the reality of community lies in its members' perception of the vitality of its culture. People construct community symbolically, making it a resource and repository of meaning and a referent of their identity" (Cohen 1985: 118). By definition, the New Media Nation has no clearly defined geographical boundaries. As a mediascape—a transcultural, transnational, global project using language, print, airwave-space, and time—it exemplifies the broadest possible boundary breaching extension of Anthony Cohen's vision of symbolically constructed community. It would be a mistake, though, to characterize such responses as merely "traditionalistic," implying that the community in question is mired in its own past and is unable to face up to present imperatives. Rather, "the past is being used here as a resource" (Cohen 1985: 99) for creating a new transcultural, mega-community with components and attributes consciously selected from the individual cultural communities.

New-old Adventurers and Voyeurs

In 2006, Survival International proudly, yet ambivalently, linked itself to the British filmmaker and television presenter, Bruce Parry, and his hit BBC show, *Tribe*, saying that Parry "has urged viewers of the program to support tribal peoples by getting involved with Survival." In the midst of its own campaign to improve and increase mass media coverage of Indigenous peoples and their concerns, Survival quoted Parry as saying, "We have so much to learn from tribal peoples and yet they themselves are frequently facing extreme difficulties or extinction. I hope that our series may have touched you in some way and I recommend your contacting Survival International to find out more about the global plight of indigenous peoples and how you may help." I have considerable respect for Survival International's efforts and activities, and especially for their media savvy, which has elicited support from prominent print and broadcast journalists. In this case, I think they may have fallen prey to celebrity flattery and overlooked the wider implications of voyeuristic video. Bruce Parry's "adventures" are other people's lives. The shows he produces are "respectful," eroticizing, and patronizing. Viewers may remember seeing unfamiliar settings and people, who occasionally speak. But Parry's remains the authoritative voice; he tells the story and shows us how "sensitive" and "courageous" he is, to participate in the ceremonies and practices of strangers.

To be fair, Survival's director Stephen Corry did express some reservations and Survival has consistently worked to make positive use of public interest, including in this instance: "Tribal peoples are every bit as fascinating as the *Tribe* series suggests. But they also urgently need help to resist the forces stacked against them—land grabbing, resource theft, introduced disease, all underpinned by racism on the part of the authorities and companies who are guilty of destroying them. Survival helps them make their voices heard. I urge anyone who is intrigued by *Tribe* to find out more about tribal peoples and how to stop them disappearing." What he missed was the potential damage of intrusion, however sensitively arranged. While decrying the dangers of bringing disease to uncontacted tribes, Parry energetically brings his crew to every selected "location." I have never seen a suggestion that the crew are checked for illness, or masked to protect their hosts from unfamiliar bacteria that may threaten their lives as much as camera crews may threaten their privacy.

Parry calls himself an "activist," "expeditioner, and television presenter." A businessman with his own TV production and travel companies, he is adept at cashing in on "extreme" sport and adventure, "reality" TV, and other trends. He once led Commando Training in the British Royal Marines and now leads television teams. In between, he worked on pop videos and feature films before developing films of his own—such as the unfortunately titled "Cannibals and Crampons," shown on BBC's Extreme Lives series. Adventure and macho showmanship take center stage. His company website boasts that: "Everyone in Endeavour is an adventurer as well as a film maker…Endeavour Productions specializes in extreme outdoor television production. The harder, the faster the more mobile the better" (Endeavour Productions 2006). It is possible that he has made some contribution toward educating a wider public about Indigenous peoples' rights and problems. It is questionable whether such representations do more good than harm, in perpetuating images of exotic Otherness and indigeneity as adventures ripe for explorer-voyeur-visitors to savor.

"For anyone who has been watching the series and witnessed Bruce drinking fresh warm animal's blood, eating rat's tail, having his septum pierced with a three-inch thorn, and having a soul-baring hallucinogenic trip from three days of eating tree bark…it is evident that Bruce is truly on the forefront of extreme travel…with an insatiable curiosity that takes him deep in to the heart of any culture. His charisma and sensitivity to the communities make him a visitor highly respected by both tribal host and western spectator alike." Charisma, maybe. Sensitivity? His manner on camera, the BBC's description of his work, and the films themselves are uncomfortably reminiscent of early displays of Indigenous people in circuses and museums. There have always been explorers, missionaries, traders, and other visitors who "went native." The main difference is in the pretence that this is advocacy based on genuine collaboration between Parry, the film crew, and their hosts. The Parry enterprise seems lacking in self-critical analysis or sufficient reflection on the two-way mirror of transcultural experience. I much prefer the irreverence of Gary Larson, who has a laugh at colonizers' expense, in a *Far Side* cartoon spotlighting cross-cultural mis-communication. The caption reads, "The Lone Ranger, long since retired, makes an unpleasant discovery." Larson pictures the Ranger in an armchair, reading an "Indian Dictionary." Inside the cartoon bubble, he says to himself, "Oh, here it is…'Kemosabe': Apache expression for a horse's rear end" (Larson 1990: 41).

A recent effort to redress past grievances in Canada, sketched below, needed a bit more of Larson's approach and a bit less of Parry's.

Arts and Cultures: Natural and Unnatural Histories

The Royal Ontario Museum (ROM) is an agency of the Government of Ontario. Created in 1912, it calls itself "Canada's largest museum of natural history and world cultures." Its history of representing minority peoples is fraught with controversy: in recent years, its managers have made efforts to improve their public image and community relations. In 2005, the ROM announced the opening of "a new permanent gallery presenting Canadian First Peoples" art and culture. More than 1,000 objects tell stories of Native cultural expressions, ranging from 10,000 year-old archaeological artifacts to contemporary Native artwork. This is the first time in 25 years that the ROM has had a permanent location for its First Peoples collection. Ken Lister, Assistant Curator in the Department of World Cultures and coordinating curator for the gallery, said, "We have taken a non-traditional approach in presenting our Native collection" by organizing the collections "by theme and by collector rather than by the more traditional culture area approach." His use of the word "non-traditional" is interesting, since he refers, not to First Peoples' traditions but apparently, to those of Euro-Canadian curatorial practices (Royal Ontario Museum 2005).

Lister explains that "older and contemporary artefacts are married within an overall theme that showcases the First Peoples collection in the context of prevailing attitudes and collection histories." One might well ask, whose "prevailing attitudes" and whose "collection histories"? Lister declares that "Native advisors have been instrumental in the development of the gallery," as if that were an earth-shaking development. Indeed it is, but that fact merely underscores how backward cultural institutions remain. The curator makes much of the "objects chosen by the ROM's Native Advisors," who "have shared their knowledge with ROM curators and exhibit planners in selecting and interpreting these objects." One wonders why they were not made curators and planners instead of "advisors." The implication is that a favor was done by inviting their (presumably unpaid) participation. It would take a far greater stretch for the ROM to bring in paid Indigenous expert curators. The implication that there are no such people perpetuates the very paternalism and colonialism the ROM directors are presumably trying to change. The very language used to describe this effort to change old habits retains the paternalistic tone: "The descriptions were written by the Advisors, giving this section completely over to the Native voice." But in reality there is no "Native voice"; there is a multiplicity of voices and cultures—as Lister surely must know. I would like to read a statement by Mathew Nuqingaq, one of the "advisors," about his choice of a *qulliq*, an Inuit seal oil lamp, instead of Lister's voice telling us that "Mathew uses the *qulliq* as a metaphor for Inuit culture itself: The qulliq is a celebration of the seal."

In what today is Canada, and crossing Indigenous borders, in North America, there are hundreds of Indigenous artists whose works are exhibited, and acknowledged, across "traditions," cultural, political, and geographical boundaries. Yet the ROM chose to launch its new gallery "representing" First Peoples with a "gallery highlight"—"the area devoted to the ROM's important G.W. Allan collection of Paul Kane's paintings and sketches, Canada's largest collection of his art." "From 1845–1848, Kane travelled extensively through Canada, northwest of Toronto, to sketch the Native Americans of the region." Paul Kane (1810–1871) was born in Ireland and spent two-and-a-half years sketching landscapes and First Peoples in difficult terrain, inspired by the US artist, George Catlin, among others. CineFocus Canada calls him "one of the first 'tourists'"—as opposed to explorer, trapper or surveyor—to travel the northern fur-trade route from the Great Lakes to the Pacific Ocean (CineFocus Canada 2007). He also authored a best-selling travel book. There is no question that the works of Paul Kane have historical and cultural significance. What I do question is the museum's decision to place these works at the center of an exhibition presumably dedicated to giving "voice" to First Peoples and to present this exhibition as a radical departure from previous museum practices. Once again, the Indigenous "advisors" are supplementary and the authoritative voice is that of a colonial outsider. The Kane works certainly belong in a Canadian museum. They do not belong in an exhibition of supposedly first-hand depictions of Indigenous people. Once again, the curators have deemed it necessary to augment or supplement, replace or displace Indigenous people.

The media release tells us that, "Themes and Collections explores the ideas of anthropology and collecting for museums. Visitors can have a glimpse of First Peoples' art and culture through the eyes of "an archaeologist, a Mohawk woman, an art collector, anthropologist, early film maker, and artist" (Royal Ontario Museum 2005). What is striking is that this string of identities designates several specialist authorities and "a Mohawk woman." It is a depressing reminder of a point that Lorna Roth made in 1995, about a 1954 article by Pierre Berton in *Maclean's* magazine. "*Plus ça change...* The magazine had published a page of photographs labeled 'The Face of the North.' Eight of the nine people pictured are identified by their professions—'nurse, surveyor, postmistress, trapper, priest, deacon, prospector, miner'...and 'Eskimo'" (Roth 1995: 39, 50). It is shocking to see an approach from half a century before repeated in a 2005 exhibition by a major museum (Royal Ontario Museum 2006)—and again, in 2007, in a magazine that ought to know better.

In 2007, *Disability Now*—an advocacy and information magazine promoting the rights of people with disabilities in Britain—published a feature on the joys of travel for people with disabilities. In it were photographs of individuals labeled "Venezuelan children"; "a fruit-seller"; "one of the porters"; "one of the guides"; and "a Kamarata Indian." It is striking that both the author and the editor missed this. The article was part of a holiday feature aimed at encouraging the magazine's readers and others to travel, and for travel companies to provide accessible holidays. And yet, in considering the rights and dignity of disabled readers and travelers, the author-photographer failed to consider the rights and dignity of Indigenous people they met on their Venezuelan tour (Davidson 2007: 25).

The ROM picked up on an interesting trend for museums of natural history to link history, anthropology, and art. In general, I think it is a positive trend, which is often used to underscore the fact that Indigenous cultures are living cultures and not relics of the past, but it has sometimes been used to marginalize Indigenous artists. The flip side of that trend is the continuing practice of separating "art" from "Indigenous art." In a chapter for *Canada North of Sixty* (Alia 1991b), I criticized the separation of "art" and "Indigenous art" implied in the housing of works by Indigenous artists in museums of natural history, rather than in art galleries. In the Canadian context, this referred to the "Group of Seven," whom many consider to be Canada's pivotal group of landscape painters. Some people found my views perplexing or offensive. In response, I offered a tongue-in-cheek alternative:

> Apparently, my complaint that First Nations artists are too often exhibited in natural history museums, instead of in galleries with other artists, perplexed some people. I have thought the matter over, and offer the following counter-proposal.
>
> Let us leave the artists of Aboriginal descent in the Natural History museums. After all, they are a part of the history of civilization. Their works tell us much about social and cultural concerns, and often express quite explicitly the traditions that nurtured and inspired the artists. In fact, we need to integrate art even more fully into our exhibition-portraits of civilization. How will we know the depth and extent of European culture without seeing Van Gogh's portraits of peasant life or Rembrandt's portraits of the elite, or Monet's serene country gardens?
>
> With this in mind, I propose that the fine collections of "Group of Seven" landscape paintings housed at the National Gallery be moved to the National Museum of Civilization. What better way to integrate the complex and magnificent strands of our cultural heritage. These remarkable works by members of an Anglo-Celtic, male, art-focused subculture are an important component of Canadian history. They should be appreciated fully, in the context of Canadian history, not marginalized in an elite and specialized gallery which separates them from other artifacts of Canadian culture.
>
> The Group of Seven artists have much to offer, within the broader sociocultural context. They made many decisions by consensus, emphasized

> community and shared experience, and organized projects to collectively display and promote their individual works. Thus, the members of this subculture placed higher value on collective action than on individual accomplishment. They developed a common aesthetic, in which love, and cultivation, of land was paramount. They have much to teach today's Canadians about the links between spirituality, aesthetics, environment, and social organization.
>
> Isolating artists denies all this. Including them gives the public the opportunity to comprehend them as a subculture which made a substantial contribution to the development of civilization in what we now call Canada. Integrating them into the displays of other Canadian arts gives us a fuller and more accurate view of our human history. Clearly, the works of these Canadian neo-Natives deserve to be seen alongside those of Canadian Aboriginal artists whose work derives from earlier traditions that provided the cultural underpinnings for their own era. Their passion shows the intensity with which their ancestors embraced their adopted land. I am certain that audiences will appreciate the Group of Seven and its production more, in the museum setting, and look to the public to lobby our museums and cultural ministries to support my proposal (Alia 1991).

Such images and texts are being challenged—and not just through letters to the editor and efforts to improve mainstream media coverage. Increasingly, Indigenous people are taking portrayals of their cultures, lives, concerns, and communities into their own hands and voices. One of my favorite artists, Jane Ash Poitras, makes gorgeous collages filled with her own heritage and the images of other Indigenous people, their histories, and symbols. In her wonderful visual stew, she challenges the nature and meanings of cultural appropriation. Her works are filled with homecomings and leavings, homelands and Diasporas. She celebrates and questions "coming home" with images of Indigenous people from everywhere.

Elitism and ethnocentrism are imbedded in art history and criticism. Mainstream art history and criticism relegates Aboriginal art and artists to margins, footnotes, and appendices. The Métis filmmaker Loretta Todd says that appropriation occurs when "someone else becomes the expert on your experience (Fung 1993: 18). Videographer Richard Fung says that the free exchange of foods, religions, languages, and other cultural attributes means that "most of what we think of as culture involves some degree of appropriation ... there are no clear boundaries where one culture ends and another begins" (Fung 1993: 17). Fung says that "appropriating someone else's voice, sound, imagery, movement or stories can represent sharing or exploitation, mutual learning or silencing, collaboration or unfair gain, and, more often than not, both aspects simultaneously." Fung's view supports the pan-indigeneity imbedded in the Fourth World movement and The New Media Nation. It involves not only cultural and geographic crossings, but disciplinary crossings as well.

The New Media Nation challenges notions of "media" and media-makers, mixing art and science, news making and reporting, religion, politics, and culture. It is an adventure and an experiment in communication, a celebration and an on-going conference with numerous stopping points and milestones. One of these, the first International Forum of United Indigenous Peoples, was held in June 2006 in Pau, France. The weeklong Forum opened with "a message of peace delivered by indigenous women from 30 different cultures on five continents. Rigoberta Menchu Tum, the Nobel Prize-winning Quiche Indian who fought for human rights in Guatemala, was among 350 delegates in attendance." Also attending the conference were Inuit and 15 North American First Nations, including a large number of Haida elders, artists, and political leaders from western Canada. "They put us kind of front and centre on it, I guess because of the fight we've been going through, and the successes in the court and on the land," said Guujaaw, president of the Council of Haida Nations. In the 1980s, the Haida negotiated an agreement setting aside about half of Moresby Island (*Gwaii Haanas*) as a protected wilderness area, "a success that caught the imagination of indigenous people around the world."

Guujaaw told CBC Radio, "Our people are all very political because that is the first priority. You make sure that the culture will survive...governments can't just push us aside...we take full advantage of that and we've been pushing back." Haida art, as well as politics, was on show at the Forum, underscoring the point that "Haida don't separate their culture from their political struggle or the survival of their people." Nika Collison, co-chair of the Haida Repatriation Committee, said there is no Haida word for art, "because it's part of everything in our life." The Forum also included a film festival with works by First Nations filmmakers and Jean Malaurie's *Last Kings of Thule*, a documentary of Indigenous life in Greenland (CBC 19 June 2006). The showcasing of Haida culture and politics sent a clear message about the importance of negotiating with particular Indigenous and state governments, along with pan-Indigenous collaboration.

In Canada, networking between and among Indigenous cultures and communities—both within and outside Canada—has been well developed and extensively utilized as an organizing tool. But things were not always this way. It has been a bumpy ride, even in the best of circumstances. Let us consider some of the historical, and current, reasons Indigenous people are "reclaiming the future."

CHAPTER 2

Pathways and Obstacles

Government Policy and Media (Mis)Representation

You've got to have a willing heart to make things right.
—Clifford George, Stoney First Nation elder

George Githii, former editor-in-chief of the *Daily Nation* in Kenya has said: "For governments which fear newspapers there is one consolation. We have known many instances here in which governments have taken over newspapers, but we have not known a single incident in which a newspaper has taken over a government" (Githii, quoted in Robie 2005: 2). The *Tundra Times* came close, for a few historic, politically pivotal moments. In 1962, it took on the US and Alaska governments. In a masterful mix of political strategizing and journalistic muckraking, it helped derail an Atomic Energy Commission plan to use an above-ground atomic blast to excavate a harbor near the Eskimo village of Port Hope, and became a major force in the successful struggle for ANCSA—the Alaska Native Claims Settlement Act of 1971 (Morgan 1988: 107-200; 222).

Referring to his home country of Canada, the broadcaster, writer, and academic, Michael Posluns, has "been struck, for many years, by the enormous effects [of] the array of civil disabilities imposed by Parliament," particularly the 1927 amendment to the Indian Act amendment, which made it illegal to raise funds "for the purpose of pressing Indian land claims," prohibited potlatches and other giveaways, and gatherings of more than three persons. In the light of such proscriptions and restrictions, Posluns thinks it "is not surprising that dramatic and strong national leadership did not emerge until a decade after" these were repealed in 1951. "Today, we have the grandchildren of the leadership of the 1960s coming into their adulthood, the first generation to achieve maturity without having experienced gross and totalitarian repression

... I see a convergence, in this generation, of access to technology and the relative lack of state repression" (Posluns 2007).

That convergence has fostered the New Media Nation. The urgency of its development can be understood not only by the suppression of political activities and cultural practices, but by the rampant misrepresentation of Indigenous people in the dominant media. Whether benignly ignorant or aggressively racist, such misrepresentation does considerable damage.

Stumbles and Blunders, Intentional and Un...

Across the United Kingdom, on 4 October 2002, viewers learned about the Queen's annual visit to Canada, where, in a startling mix of half-truths and total misinformation, the audience saw pictures of "Baffin Island", where Her Majesty "was greeted by some of the local Inuit Indians." Baffin is Canada's largest island and the size of many a small country. The Queen landed in Iqaluit, since 1999, the capital of Nunavut Territory. Nunavut (which the news report did not even mention) covers nearly two million square kilometers—about one-fifth of Canada's land mass (e.g., ITV 2002).

Misidentifying the place was a minor blunder compared with the misidentification of people. There is no such thing as "Inuit Indians." Inuit are people whom outsiders used to call "Eskimos." In Canada and the US, some non-Inuit Indigenous people are called "Indians"—an error dating back to Columbus's arrival in 1492 and erroneous assumption that he had discovered India. North American "Indians"—who are entirely unrelated to Inuit—are variously referred to as Native Americans, First Nations, Indigenous or Aboriginal people, or by culturally specific names such as Ojibway, Cree, Tlingit, Gwich'in, Delaware. Except for an occasional immigrant to the region, none of these "Indian" people live in Nunavut. "Inuit Indians" is something the writer made up and the broadcaster extended to the airwaves, thereby misinforming thousands.

A 2003 science article in *The Independent* (Burne 2003: 8) refers to "Inuits"—the equivalent of calling (male) residents of England 'Englishmans.' Any reader knows that is absurd; any editor would correct the error. In the case of 'Inuit' (which is the correct plural form—the singular being 'Inuk')", the editor missed the error. However unintentional, such carelessness borders on racism; such mistakes tend to be made about minority people in faraway places. A reporter assigned to an unfamiliar place is supposed to research the facts and the place. Imagine a newscast from London's Heathrow Airport, saying: "The Queen visited Northumberland and was greeted by some of the local English Italians." This was far from an isolated incident. In a letter to *The Independent on Sunday*, which the editor declined to publish, I wrote:

> Jeremy Seal's interesting cover story ... 'Why Santa went to the North Pole', includes an unfortunate error. It refers to 'a pair of Eskimo Indians exhibited at Tavistock, Devon.' ... 'Eskimos'—who generally call themselves Inuit—have no connection either to the people of India or to the Native American and First Nations people whom Columbus misidentified as 'Indians.' It is lamentable that key reference works such as the *Oxford Dictionary for Writers and Editors* provide information so stunningly wrong it is sometimes hard to untangle ... Surely it is time to set the record straight (Alia 2004: 55–59).

As Edward Said notes,

> We are beginning to learn that de-colonisation was not the termination of imperial relationships but merely the extending of a geo-political web which has been spinning since the Renaissance. The new media have the power to penetrate more deeply into a "receiving" culture than any previous manifestation of Western technology (Said 1994: 353).

Media depictions of Indigenous peoples are steeped in the language of conquest and colonization. Despite centuries of change and decades of progress, that language persists. Explorers are depicted as quasi-military conquerors who launch "assaults" on remote regions and unknown lands. Indigenous people are seen as aiding in those assaults and increasing the comforts of the conquerors, ignored, or treated as exotic items for study or observation, in need of "civilization." Sometimes, as in the case of Sherpas, who have long aided non-Indigenous mountain climbers, their very names go into generic usage. In 1991, Britain's *Daily Telegraph* sent a team for a brief visit to Holman Island, documented in a photo-essay headlined "Dressed to Kill: Hunting with the Eskimos of Holman Island." It angered members of the Holman (Uluqsaqtuuq) community and many others. In a letter to the editor, which *The Telegraph* did not publish, Holman Mayor Gary Bristow, Community Corporation Chairman Robert Kuptana, and anthropologist Richard Condon wrote:

> thousands of ... readers have had their opinions and attitudes about the Canadian Arctic falsely influenced by individuals with no understanding of even the most basic aspects of Canadian Inuit culture (Condon 1992; Bristow, Condon and Kuptana 1992).

Among the most "grotesque" of *The Telegraph*'s errors:

> Among hunters there is no code of honour, the article proclaims ... apparently measuring Inuit hunting against aristocratic English fox hunts. 'The hunter ... is merciless and self-interested, gathering food only for himself and his family ... The unsuccessful hunter and his family could go hungry only steps away from someone else's well-stocked tent (Thompson1992: A3).

The Telegraph did not take the trouble to research Inuit hunting and food sharing practices. The *Telegraph* has telegraphed a portrait of "wild" Indigenous people much like the "wild Indians" of older racist discourse, brought back to life in the fashion catalogue described below. No major newspaper would publish photographs of majority-culture people without identifying them. Yet two of Holman's most respected elders, Jimmy and Nora Memogama, are pictured without identification.The most amusing blunder is the *Telegraph's* tale of a "young white man who stepped off a train to stretch his legs; his frozen body was discovered the following spring" (*Telegraph Magazine* 1991). It is possible that, realizing the journalist's ignorance and gullibility, someone had a laugh at his expense. Nevertheless, his report went into the newspaper, and thus, the historical record. No one has ever stepped off a train at Holman. The nearest railhead is more than a thousand miles away, near the Alberta border (Bristow, Kuptana, and Condon 1992). I tell this story often, and it always gets a laugh. But this is not just an amusing example of journalistic ignorance. It is:

> representative of a disturbing trend in journalistic coverage ... Each year ... Holman is visited by journalists who desire to write or photograph the definitive article about an isolated Inuit community ... the community has no way to ... monitor or comment upon their finished works ... [the] worst harm is not the offence they give to ... residents but the distorted view they present to thousands of readers about northern life and northern people. In an age when Inuit culture is being attacked by numerous animal rights groups, articles like "Dressed to Kill" ... perpetuate prejudice (Bristow, Kuptana, and Condon 1992).

Accuracy is a basic tenet of journalistic practice. Having failed to check the facts and having appended a headline that amounts to hate literature, the editor exacerbated the damage by refusing to publish the letter from Holman residents, correcting the errors.

While still having to address current and continuing media malpractice, Indigenous people have centuries of media problems to correct.

In the early twentieth century, advertisements and films were filled with images of "'Indian princesses' and other imaginary people. 'Indian' is a term used to sell things—souvenirs...cigarettes... cars...movies...books. 'Indian' is a figment of the white man's imagination," says the Ojibway storyteller and writer, Lenore Keeshig-Tobias (1990). The woman known as Sakajawea or Sakakawea inspired a 1920s cleaning and dyeing advertisement of an "'ageless' and shapely Indian princess with perfect Caucasian features, dressed in a tight-fitting red tunic, spearing fish with a bow and arrow from a birch bark canoe suspended on a mountain-rimmed, moonlit lake." Another Aboriginal damsel was used to promote Chippewa's Pride Beer (Valaskakis 1995: 11). Things have not improved in the twenty-first century.

In 2002, an ash-blond boy in a generic headdress promotes a Japanese car with "ample space for *braves* and *big chiefs...young Hiawatha*...needn't leave *his headdress* behind, which could avoid some ruffled *feathers*." The puns suggest fun, presumably at no one's expense (Suzuki 2002). As we will see in the discussion of Ainu history and society, Ainu are challenging such attitudes with regard to not only depictions of other Indigenous peoples, but also their own relationship to Japan's dominant society.

The spring 2003 catalogue of a Swedish designer—who ironically is proud of her Sámi heritage—says that her collection was inspired by "wild, colourful Native Americans."

> Squaw, Pawnee & a little Kickapoo. Sometimes you just feel like doing something wild ... my rebellious artistic soul ... loves the unexpected and surprising ... And so the Native American theme was born ... It's so exciting, don't you think? Yee-haa! (Sjödén 2003)

Fake "Native American" feathers combine with fringes, beads, and geometric patterns. The text is presented in the designer's voice; footnotes add an academic air; the multilayered misinformation is difficult to unpeel. "Red Indians prefer to be called 'Native Americans' ... Chippewa is [the] highest leader (Chief) in the Algonquian tribe ... Squaw [is] A Native American woman." A blonde model with feathers in her hair, identified as "Squaw Hanna," "dances wild in a native American-inspired tunic ... and pretty ikat-weave silk scarf." Ikat is an *Indonesian* weaving technique; 'squaw' is a term that many Native Americans and others find offensive; "red Indians" is a pejorative whose persistence in current usage is hard to comprehend (Sjödén 2003).

In 2005, the Australian Broadcasting Corporation, among many other media outlets, reported the discovery of "Australia's earliest human footprints." It was a spectacular find.

> At tens of thousands of years old the find is the largest group of human footprints from the Pleistocene era ever found. Archaeologist Dr Matthew Cupper of the University of Melbourne and colleagues report their findings from the New South Wales Willandra Lakes World Heritage Area online ahead of print publication in the *Journal of Human Evolution.* "It's a little snapshot in time", says Cupper. "The possibilities are endless in terms of getting a window into past Aboriginal society." (Salleh 2005)

The manner of reporting about the "window into past Aboriginal society" keeps the windows shut against understanding the present—and *presence*—of Aboriginal society. The report not only gives short shrift to living cultures and communities, it fosters the pretence that Matthew Cupper is the 'discoverer.' The real discoverer was an Indigenous archaeology student named Mary Pappin. Cupper's account, and the journalist's report, patronize and trivialize her role:

> Cupper says a young woman by the name of Mary Pappin Jnr of the Mutthi Mutthi people found the footprints in August 2003 while exploring the area with team member Professor Steve Webb of Bond University on Queensland's Gold Coast, as part of a project to educate young Aboriginal people in archaeology (Salleh 2005).

"It's quite remarkable," says Cupper. "We haven't found any footprints from the Pleistocene in Australia before ... the prints are also the largest group of Pleistocene human footprints in the world" (Salleh 2005). And who, we might ask, are the "We" in Cupper's account? He seems clearly to be referring to the "scientific community," of which Indigenous people are "students" rather than members. The distinction is carried further with the implied distinction between (rational) "science"and Indigenous "emotion."

> It's quite an element of excitement with the local Aboriginal community. It is a really strong emotive link with some of their past ancestors', he says. Executive officer of the Willandra Lakes World Heritage Area, Michael Westaway, told the ABC that the original discoverer of the footprints Mary Pappin Jnr was unavailable for comment (Salleh 2005).

"Indigenous peoples must either obtain equal access to the political, cultural and economic opportunities of the colonising society, or continue to struggle for political autonomy. The media play an important role in that struggle" (Robie 2005: 1).

The Roles of Government

> While government policy ... has run the gamut of genocide, expulsion, segregation, and assimilation, the one constant is that governments have never 'genuinely recognized Aboriginal peoples as distinct Peoples with cultures different from, but not inferior to, their own.'
> –Will Kymlicka

> Alone, you can fight
> ...
> But two people fighting
> back to back can cut through
> a mob...
> Three people are a delegation,
> a committee, a wedge. With four
> you can...start
> an organization...
> A dozen make a demonstration.
> A hundred fill a hall.
> A thousand have solidarity and your own newsletter;
> ten thousand, power and your own paper;
> a hundred thousand, your own media;
> ten million, your own country.
> ...it starts when you care
> to act...
> when you say We
> and know who you mean, and each
> day you mean one more
> —Marge Piercy

The New Media Nation is an ever-expanding global community composed of people who know who "We" are. Indigenous people are finding ways to transcend the limitations imposed by the governments of the particular countries in which they reside. Nevertheless, the progress of global media networks is deeply affected by the policies, politics, and priorities of those governments. The communications policies of different countries have variously restricted and enabled the development of Indigenous media. For example, Canada's multiculturalism and commitment to serving remote communities has fostered an array of media projects, experiments, and services, while the long-standing and entrenched assimilationist policies of the United States have had very different results. There is far less public support of ethnic minority and Indigenous cultural and media projects, and little interest in serving the needs of remote and otherwise disadvantaged communities. Indigenous people have been among the beneficiaries of Canada's broad support of multicultural institutions and minority media. The result is a multiplicity of public voices, which, to my knowledge, is unique in the world. Such policies are not without their costs. With a focus on creating communications outlets and infrastructures, the development of transport came second. One factor is the role of the United States in setting transportation priorities—its keenness to develop transport networks in northern Canada increased during the Second World War and the Cold War.

Although its Conservative government tried to reverse the pattern by refusing to sign the Universal Declaration of the Rights of Indigenous Peoples, Canada has long involved itself in language and minority rights, and particularly in communications—thus, the distinction of its National Film Board, the Canadian Broadcasting Corporation (CBC), the Aboriginal Communications Societies, and APTN. While Canada has a longstanding alternative tradition of what Liora Salter calls "constituency-based services," it has shifted from a regionally inclusive "right-to-receive services" approach to enshrining the rights of women, First Nations peoples, and multicultural and multiracial ('visible minority') communities. The 1991 Broadcasting Act specifies the right of all Canadians to be fairly portrayed and equitably hired in public, private, and community broadcast media. Except for medical evacuation and other subsidized emergencies, northern air travel remains too costly for many, while an elaborate network of communication structures enables people to 'visit' in low-cost broadcast space or cyberspace from home, or in community centers with public computer access. By contrast, the Soviet Union invested little in Indigenous peoples or regional communications and instead invested greatly in transport networks in its remote and northern regions; though in recent years, much of the transportation network has been undermined by the decline of government support. Below is a sketch of some of the ways in which government policies have affected Indigenous peoples and the evolution of their political and media projects.

The Russian Federation

In 1989, Mikhail Gorbachev formally recognized the Association of Indigenous Numerically Small peoples of the Russian Federation North, Siberia, and the Far East (Russian acronym, AKMNS), now known as the Russian Association of Indigenous Peoples of the North (with the English acronym, RAIPON). It was formally founded in 1990, at the first conference of Indigenous Peoples of the North. Its member organizations were: Yasavei (Nenets Autonomous Area), Yamal for Future Generations (Yamal-Nenets Autonomous Area), Saving Yugra (Khanty-Mansi Autonomous Area), Association of Kola Sámi (Murmansk), Selkup Kolta Kup (Tomsk), Evenk Arun (Evenk Autonomous Area), peoples of Yakutia (Dolgan, Evenk, Even, Yukagir, Chukchi), Taimyr peoples (Dolgan, N'ganasan, Enetse), peoples of Sakhalin (Nivkh, Orok), Evenk of Chita, Oroches and Nanai of Khabarovsk Territory, Itelmen of Kamchatka, and native peoples of Chukotka and Kolyma. RAIPON represents over 190,000 members. It is the only organization empowered to represent the 30 Indigenous peoples of the Russian northern regions.

Its mandate includes protection of rights, development of publishing houses, support for languages and cultures, international cooperation, and exchange. Its first president, Vladimir Mikhailovich Sangi, the Nivkh poet and novelist, was born in 1935 on the east coast of Sakhalin Island in the Nabil nomad camp. In 1991, I met him in Alaska and interviewed him, with the help of the anthropologist, Igor Krupnik, who served as a Russian language interpreter. Sangi had "founded Siberia's Aboriginal people's movement" more than thirty years earlier. His hope was to save the people of the North by giving "the power to the people themselves." He expected that "this movement of Arctic minorities—such small pieces of ethnic groups—[would] stimulate the whole destruction [of] this ugly federation we have." His goal was to destroy the system and "to create within the borders of Russia a state of equal peoples" (Alia 1995: 21). While that goal is unlikely to be realized in the foreseeable future, the years that followed our interview saw an increase in attention to Indigenous rights and cultures. Yet government and corporate interests continue to exploit the resources and people of Russia's northern regions, while efforts continue to improve education, retention, and restoration of Indigenous languages and cultures.

In 1999, RAIPON held an international round table on an Indigenous parliament. In 2002, there was a joint session of the RAIPON

Coordination Council and the National Organizing Committee for the International Decade of the World's Indigenous Peoples in Yakutsk. Among other things, the law of Sakha, *On the Legal Status of Indigenous Peoples of the North*, guarantees the creation of a climate aimed at protecting the cultures, economic, political, and social development of Indigenous peoples (Zakharov 2005: 88).

Despite the rampant health, economic, and environmental problems experienced by so many, Indigenous peoples are experiencing a cultural renaissance that parallels those of Māori in Aotearoa, Inuit in Canada, and peoples in other parts of the world. For example, the Evens (Eveny) of northeast Siberia broadcast in the Even language from the Gyavan radio station in Yakutsk; there are Even language newspapers in Yakutia and Chukotka. The language is taught in secondary schools, at Yakut State University, and at the Herzen Teacher Training Institute in St. Petersburg. They are reviving traditional holidays, music, and dance (Golubchikova, Khvtisiashvili, and Akbalyan 2005).

In a 2006 report for the BBC, Jorn Madslien said the "traditional way of life has been under assault for decades," with Sámi "gradually forced off arctic Russia's fertile tundra grazing-land and into artificially created towns." Contact between Russian and Nordic Sámi was curtailed by closure of the border during the Cold War years. "Much of the displacement was caused by a steady expansion of industry, forestry and mining, and the arrival of hundreds of thousands of workers from other parts of the Soviet Union—many of them … forced labourers in Gulag camps." Sámi coastal fishermen "were ordered to move away from the shores of the Barents Sea … currently littered with secretive navy installations, and reindeer herders were forced away from a 200-mile exclusion zone that ran along the Cold War frontier" (Madslien 2006, unpaginated).

Vatonena Lyubov, vice president of the Association of Kola Sámi, has said, "We will never regain our grazing lands and our rivers." The idea of private ownership of land and the pressure to compete were new to Sámi, who now face rampant unemployment. They also live with health risks caused by mining, smelting, oil and gas exploration, and a proliferation of nuclear power plants and nuclear waste. The oil and gas industry threatens to encroach on their territory (Golubchikova, Khvtisiashvili, and Akbalyan 2005, unpaginated). At Lovozero, a 'concrete jungle,' the Sámi "flag is flying proudly over a cultural centre" built in 2003, where the once prohibited Sámi language "is spoken on the airwaves via the nation's own radio studios. This is mission control for those who want to preserve Sami traditions and culture." Sámi organized a public protest in 1998, against efforts by a Swedish company to dig an open-pit gold mine "in the heart of the grazing lands." Some Sámi now wear traditional garments once banned by the Soviets, which they have recreated using archival photographs. They are demanding a share in mineral and other resource rights and profits (Madslien 2006).

Across the Russian North, there are 41 Indigenous cultures (Wessendorf 2005: 9). Among Russia's contemporary Indigenous media practitioners, Galina Diatchkova highlights the work of two Chukchi journalists, Elena Timonina and Margarita Belichenko. Belichenko is a producer of Chukchi language radio programs. Timonina is a television journalist based in Anadyr City, Chukotka, who has produced an array of Chukchi-language television programming and, in collaboration with Alexander Rudoy, the film, *When the Men Cry,* about a Chukchi boy learning to herd reindeer (Diatchkova 2008). Publications preceded broadcasting in this region, and generally have had a stronger presence. The first books and periodicals in Indigenous languages were published in the early nineteenth century by the Russian Bible Society, and were mainly religious texts. Most secular books came later. For all their cultural diversity, the Indigenous writers and journalists share what seems a very Russian literary tradition, in that so many have been both journalists and poets. Here are some of the leaders, and leading projects, in Indigenous media (Golubchikova, Khvtisiashvili, and Akbalyan 2005: unpaginated online edition):

> Anempodist Ivanovich Sofronov (1886–1935) wrote in the Yakut language and is considered a pioneer of Yakut literature. He worked as a typesetter for the newspaper *Yakutski Krai,* wrote for the magazine *Golos Yakuta,* and founded a national theater for which he wrote plays. He was also a songwriter and poet, editor-in-chief of *Manchary,* the first Soviet newspaper published in the Yakut language, and editor of the first literary magazine, *Chalbon* (*Morning Star*).
>
> Ivan Yakovlevich Babtsev (1892–1962), was an Evenk writer, playwright, and scholar of Evenk language and folklore. In the late 1930s, he recorded Evenk fairy tales, some of which were published in the Kolyma Evenk newspaper, *Orotty Pravda.* He was born in the nomadic camp of Lankovaya, lived in the Kolyma area of what is now Magadan, and helped found the first collective reindeer herding farms.
>
> Aleksei Andreyevich Ivanov (Kyunde) (1898–1934) was a Yakut poet, writing in the Yakut language, born in Kangalassy Nasleg, Yakutia. He worked as teacher and in 1926, published a Yakut grammar. In the late 1920s, he edited the Yakut newspaper. *Kyym.*

The Khanty writer and journalist, Grigori Dmitrievich Lazarev (1917–1979), wrote in the Khanty language. He worked as an editor in Lariak (now Nizhnevartovsk) District and in 1957, founded and edited the first area newspaper published in the Khanty language.

Mkar Ivanovich Kuzmin (Makar Khara) (1915–1981) wrote in his native Yakut language. He studied at Yakutsk Agricultural College and the Moscow Communist Institute of Journalism, wrote for local newspapers and published several volumes of poetry.

The multilingual Nenets author, Ivan Grigoryevich (1917–1988), wrote poetry, prose, drama, and journalism in Russian, Komi, and Nenets and was also an artist. Born in Yamal, he taught Russian and Nenets languages, graphics, and drawing at Salkhard School of Political Education. He was a co-founder of the Salekhard literary circle and of the magazine, *Isry Yamala* (*Sparks of Yamal*). His first book of poetry in the Nenets language, *Our North*, was published in Leningrad in 1953. The novel *Zhivun* (written in the Komi language) is considered his most important work.

Feodosi Semenovich Donskoi (born in 1919) is a journalist and scholar who founded RAIPON's economic-sociological research program. He was a staff reporter for the newspaper, *Sotsialisticheskaya Yakutia*, executive editor of *Eder Kommunist* newspaper, and editor in chief of the *Yakutia Republic Radiocommittee.*

The Khanty poet, Vladimir Semenovich Voldin (1938–1971), wrote in the Khanty language and was a correspondent for area radio.

Vasili Kondratyevich Alymov (born in 1883) was a Sámi researcher and journalist. In 1927, he developed a reindeer breeding project on the Kola Peninsula, and from the late 1920s to early 1930s, helped to found newspapers and magazines for Soviet Sámi.

Yuri Konstantinovich Vasyutov (b. 1931), Komi poet, writes in the Komi language and has worked as a radio operator for a logging project and as a correspondent for local newspapers.

Tasyan Mikhailovich Tein (b. 1938), Eskimo writer and archaeologist, comes from Naukan village on Chukotka Peninsula, has worked as a radio technician and teacher and has written children's songs and stories.

Natalya Nikolaevna Selivanova (b. 1943) is Itelmen and writes in Russian. Born in Sobolevo Village in Kamchatka, she was a journalist for the newspaper *Kamchatskaya Pravda* and has published short stories and other works.

The Khanty poet, Prokopi Yermolaevich Saltykov (1934–1994), wrote in the Khanty language. He was born in the nomadic camp of Nartygort on the Yamal Peninsula, worked as a teacher, and was an editor of Khanty language radio programs.

Nikolai Anisimovich Popov (b. 1929) is Dolgan, born on the Avamskaya tundra. He was a teacher and a journalist with the Taimyr region radio station and the newspaper *Sovetski Taimyr,* and has written for the newspapers *Tikhookeanskaya Zvezda, Krasnoyarski Rabochi*, and *Krasnoyarski Komsomolets.*

The Nenets poet, journalist, and broadcaster, Lubov Prokopyevna Nenyang (Komarova) (1931–1996), wrote in the Nenets language. He was a correspondent for *Sovetski Taimyr* newspaper and editor of Nenets broadcasts on *Taimyr Radio.*

The Eskimo poet, Zoya Nikolaeva Nenlumkina (b. 1950), writes in the Eskimo language. Born in Naukan Village in Chukotka, she graduated from Anadyr Teachers College. During her student years, she was heard in Eskimo language broadcasts on Chukotka radio. Her first collection of poetry, *Birds of Naukan*, was published in Eskimo and Russian in 1979.

Alitet (Albert) Nikolaevich Nemtushkin (b. 1939) is Evenk, writing in the Evenk language, and has written for the newspaper *Sovetskaya Evenkia.*

Leonid Vasilyevich Laptsui (1932–1982), a Nenets poet writing in the Nenets language, worked as a reindeer herder and fisher before graduating from Salkhard Medical School and the Journalism Department of Komsomol secondary school of the Communist Party School. He edited the newspaper *Nariana Ngerm* (*Red North*) for fifteen years. His poetry was originally published in Nenets and later translated into Russian.

> Tatiana Yuryevna Achirgina (b. 1944) is an Eskimo poet writing in Russian. Born in village of Markovo in Chukotka, into the family of Russia's first Eskimo lawyer, she studied journalism at Ural University.
>
> The Chukchi poet, Aleksandr Ataukai (1932–1974), wrote in Chukchi. Born Alkatvaam Village of Bering District, Chukotka, he published his first poems in the newspaper, *Sovetskaya Chukotka.*
>
> Vasili Spiridonovich Keimetinov, is an Evenk poet and collector of folklore, writing in Evenk. He was born in the village of Sebyan-Kyuel in Yakutia. In the 1980s, he published a monthly Evenk language supplement to the Tompo District newspaper, *Krasnoe Znamya* (*Red Banner*).
>
> During the Soviet period, the government developed departments for Indigenous peoples in schools and universities and published folklore and language textbooks. Some teachers were inspired and/or forced by the lack of materials, to write their own teaching texts. For example, in 1930, N. Tkachuk, a teacher in the Arka settlement on the Okhotsk coast, wrote a textbook in the Even language. A year later (1931), a group of Koryak students compiled an alphabet book for local use and the USSR Central Publishers produced an alphabet text for Shuryshkarsi Khant people, using graphics by the Nenets teacher P.E. Khatanzeyev, and the *Initial Tungus Book* by G.M. Vasilevich, in the Evenk language.

In 1932, the first All-Russian Conference on the Development of the Language and Letters of People of the North was held in Leningrad. Its results included publication of textbooks in 14 Indigenous languages over three years, including Chukchi, Nanais, Nivkh, Udegey, Itelmen, Inuktitut, and Even (in 1932); Mansis, Selkup, Evenk, Khant, and Sámi (in 1922); and Kets, in 1934. The first works by Indigenous writers appeared in the mid 1930s. Between 1945 and 1972, 344 books were published in eleven languages. From 1981 and 1985, the Leningrad division of Prosveshchenie Publishers issued 44 textbooks, 11 books "for supplemental reading," 10 bilingual dictionaries, four teacher training texts, and nine "aid books for teachers" in 12 Indigenous languages. Between 1986 and 1990, more than 90 textbooks were published in 14 languages, along with bilingual dictionaries for Sámi and Selkup peoples.

Today, Siberia is experiencing a crisis, due largely to changes in government and the shift away from subsidized transport. Where air travel over vast distances, at bus fare prices, once enabled indigenous and other northerners to make frequent visits among remote communities, the Russian government can no longer subsidize air travel at such a level. "Cheap aviation used to hold communities together, and now it's a mess" (Vitebsky 2001). In 2000, Roman Abramovich replaced the former governor of Chukotka, Alexander Nazarov, who was "widely despised for trying to halt the subsistence practices of Chukotka's indigenous people" and, according to Johnson, "did everything he could to break down native organizations"'(Nunatsiaq News 2002: 2). The monthly newspaper, *Aborigen Kamchatki*, "serves all of Kamchatka's aboriginal peoples ..." Published in Russian, with "occasional snippets of aboriginal language," it receives funding from The Sacred Earth, an ecotourism and environmental lobbying group based in the United States (George 2001: 13). Edited for many years by Valentina Uspenskaya, its current Deputy Chief Editor is A.V. Morilova. It functions only in print, though a brief description and Russian language contact address are available online. It describes itself as "the printed voice of the Association of Indigenous Peoples of the Kamchatkan Oblast ... not only aimed at the indigenous population and 'old settlers' of the peninsula, but ... everybody who ... values the original cultures, customs, traditions and ceremonies of the Koryak, Itelmen, Even, Chukchi, Aleut and other peoples of Kamchatka" (Morilova 2007). Articles cover "the socio-economic situation of the indigenous peoples, activities connected with their traditional livelihoods, history, culture, art, ethnography, health, education, and language revival of the Northern peoples." Mainly in Russian, it publishes some material in Koryak, Itelmen, Aleut, and Even languages, as part of its mandate to promote language retention and learning. It is distributed by subscription and provided free of charge to indigenous peoples' organizations of the Kamchatkan Oblast, the Koryak Autonomous Okrug, and Russia (Morilova 2007).There is additional support from new and continuing programs at Kamchatka State University. Its 2007 website features two programs developed by and for Indigenous people. The International Center of Traditional Cultures of Kamchatka Natives "was created in order to promote preservation and development of cultures of Kamchatka natives, unification and coordination of scientific research work and cultural activity of Russian and foreign specialists in this sphere." Together with foreign partners, it conducts research on Indigenous history and cultures, along with more activist social, economic, and cultural projects, and organizes international conferences, seminars, expeditions, and festivals in Kamchatka. "Post-graduate courses function at the center. Themes of post-graduates' theses are devoted to different aspects of Northerners' development" (Kamchatka State University 2007).

The second program is the North People Academy (NPA), established in 2000 at the university's Institute of Regional Human Problems (KSTTU). It "comprises Schools of Kamchatka native languages (Itelmen, Koryak, Aleut), School of Native Handicrafts, Kayur School, School of Traditional Technologies and Land Use, Master-class 'Rhythms of the North.'" It is staffed by Indigenous instructors who "seek to revive and preserve language and culture of their ancestors." One of them, the journalist and academic, Valentina I. Uspenskaya, former editor of *Aborigen Kamchatki*, is a key member of the NPA staff. The focus is on community education. NPA provides "regular access for children and adults, students and instructors to traditional knowledge and culture of Kamchatka indigenous peoples." Linking "science, life, policy and art," its programs are "open to one and all" (Kamchatka State University 2007).

Joachim Otto Habeck considers Internet communications essential for improving the situation of Indigenous peoples in the Russian Federation. Nearly a decade after his observations, such resources remain scarce and are produced and maintained mainly by those outside the communities they represent. The main exception is Koryak Net, "one of the very few Web resources focusing on one distinct indigenous people in the Russian Federation." In addition, a 1995 article by Kola Sámi scholar, Yelena Sergeyevna, "was edited as a Web page and went on-line in 1995 or 1996." Even so, the page is sponsored from outside Russia, as part of a service by the Sámi Association of North America (Habeck 1998: 278).

Figure 2.1. Broadcast facilities at Lovozero, Russian Federation, on the Kola Peninsula. Photograph by Jorn Madslien.

Figure 2.2. Evgeni Kirillov, producer, in the Sámi Radio studio at Lovozero. Photo by Jorn Madslien.

Australia

Australia's policies on Indigenous media have varied and shifted over the years, reflecting differences over the issues raised by Kymlicka (above) and by Murphy and Brigg (below). There are often tensions between federal and regional governments. In 2005, Northern Territory administrator Ted Egan went head-to-head with Federal Indigenous Affairs Minister Amanda Vanstone, over the provision of Aboriginal out-stations.

> The Northern Territory administrator has launched a spirited defence of Aboriginal out-stations ... Federal Indigenous Affairs Minister Amanda Vanstone has labelled small Indigenous out-stations "cultural museums" and says it is not always economical to provide them with schools, water and sewerage (ABC December 2005).

Ted Egan, "one of the first proponents of the out-station movement," emphasizes both the importance and the affordability of out-stations. He says Aborigines who choose to live on their land need only minimal services such as housing and water. "[A] community of 20 people in the middle of Arnhem Land ... wouldn't want flash towns ... just ... the ongoing acknowledgment that they have got the right to exist and be a presence on their homeland" (Egan, quoted in ABC December 2005).

> The debate about policy failure in Aboriginal affairs ignores the cultural assumptions and biases of a white worldview of "civil society" or, in the case of the history wars, the contours and constraints of white history grappling with its own story. These assumptions sustain relationships between polity and institutions that reinforce liberal Western culture ... [It] is time that all bets were off, that *everything* was up for consideration (Murphy and Brigg 2003: 3).

Thus, Lyndon Murphy and Morgan Brigg seek a total rethink of policies and relationships because, in their view, the prevailing focus:

> on administrative content and process to the exclusion of structures and values has created false oppositions in politicking about Aboriginal policy. The massive failing in all this is that the management paradigm can only evaluate its projects self-referentially, it has no way of engaging the interface between the two cultures. When governments encounter policy failure their immediate response is to modify the inputs...
>
> This repackaging only enables government to change how they do things, not what they are doing ... to reinvent new strategies for old ideas over and over (Murphy and Brigg 2003: 2).

They call for a critical examination of Australian policy and administrative programs and their "intersection with Aboriginal culture," and a "dialogue between European and Aboriginal political values and systems. It's time for conversations rather than conversion, brow beating and false oppositions" (Murphy and Brigg 2003: 3). The final years of the John Howard government featured a complex mix of detrimental, sometimes racist, policies and projects, and programs that looked distinctly forward and relatively progressive.

In 2005, Acting Race Discrimination Commissioner Tom Calma and the Attorney General, The Honorable Philip Ruddock MP, launched the 'Voices of Australia' project in Canberra to mark the thirtieth anniversary of the Racial Discrimination Act—Australia's first anti-discrimination law. Mr. Calma said, "Laws alone are not enough to eradicate racism in our society. We also need to understand one another and learn from each other's experiences so we can enjoy the great benefits of our cultural diversity. One way of doing this is to share and listen to each other's stories." The 'Voices of Australia' project published and broadcast 'real-life' stories about diversity and cooperative multi- and transcultural living in contemporary Australia, along with information about the Racial Discrimination Act, in a range of print, audio, and website resources. Audio interviews were compiled on a CD and available for download from the Internet. The Commission also produced complementary education resources, which were distributed to primary and secondary schools, community groups, libraries, local councils, religious groups, government agencies, and the general public (ABC 2005). In the 2007 national election, Australia experienced a major political shift, with the defeat of John Howard's Conservative/right government and election of Labor candidate, Kevin Rudd. Rudd arrived on a platform full of pledges, with Indigenous people front and center. The extent to which the promises will be fulfilled remains to be seen. Early reports from Indigenous and "mainstream" media were highly variable, with the optimistic and pessimistic views crossing cultural and geographic lines. Some Indigenous people had supported the Howard government and opposed Rudd. Just before Rudd's win was announced, Indigenous leader Noel Pearson accused Rudd of "abandoning his pledge to recognize Indigenous Australians in the constitution," because Rudd had told *The Australian* newspaper that "a referendum on Aboriginal reconciliation would not happen in the first term of a Rudd Labor government, if at all." Pearson said, "If you harbour any hope that these buggers are going to do anything courageous in relation to Indigenous affairs, then you're living in an illusion" (ABC 2007). He was mistaken.

Other Indigenous leaders welcomed the new government and the departure of former Minister for Indigenous Affairs Mal Brough, who lost his Queensland seat in Parliament to the Labor candidate. Brough, the "architect of the government's dramatic and controversial intervention into Northern Territory Indigenous communities, was a divisive figure. His approach was supported by such high-profile Aboriginal leaders as Galarrwuy Yunupingu and the same Noel Pearson who predicted Rudd would not apologize, but others deemed it racist, draconian and unworkable" (Gartrell 2007). Olga Havnen, CEO of the Combined Aboriginal Organizations of the NT and spokeswoman for the National Aboriginal Alliance, said, Mal Brough "has lost the trust of Aboriginal people and John Howard has lost the trust of the Australian people." Eileen Cummings, former policy adviser to the Northern Territory Chief Minister, called the election result a "moral victory" for Australia. "Aboriginal people have supported the Labor Party." Australians for Native Title and Reconciliation national director Gary Highland said the change of government could start "a new era" for Indigenous Affairs. "But we certainly can't take anything for granted and we'll still need to effectively make the case for change to achieve the sorts of things that we want." The Indigenous Tasmanian leader, Michael Mansell, said that Indigenous people across Australia "would be relieved to see the backs of Mr. Howard and Mr. Brough" (Gartrell 2007). The *National Indigenous Times* was cautiously hopeful:

> We know what Howard has been like as Prime Minister...But what do we know about Kevin Rudd? Whatever the answers are, the only thing that can be certain is that a Howard defeat will be a victory for Indigenous Australia and that in itself is a cause for celebration.
>
> On Indigenous issues, such as the Northern Territory intervention, Rudd has been extremely careful not to be wedged by the government... [He said that] a referendum on Aboriginal reconciliation would not be on the list of priorities [and] ruled out the possibility of a treaty [but] told The Australian he is committed to continuing with the controversial Northern Territory intervention [and to] making a difference on those areas of disparity between Aboriginal ad non-Aboriginal Australia (McQuire 2007).

In the United Kingdom, just after the election, *The Telegraph* reported that

> Kevin Rudd has promised to apologise to Aborigines for historic injustices ... a radical departure from his predecessor John Howard, who during 11 years in power argued that contemporary Australians bore no responsibility for past wrongs.
>
> It would be the first time that an Australian federal government had apologized to the country's 450,000 Aborigines, who after 220 years of white settlement suffer low life expectancy, poor health and high rates of joblessness and incarceration (*The Telegraph* 2007).

The Australian newspaper, *The Age*, reported that Kevin Rudd "has sketched out his vision for a more compassionate Australia, vowing to deliver a better deal for workers, the homeless and indigenous people ... [and] to close the gap between indigenous and non-indigenous life expectancy" (Schubert 2007: 1). "Seeking to allay the fears of indigenous leaders about an apology to the Stolen Generations, Mr. Rudd also confirmed that he would use the word 'sorry' and would consult indigenous communities 'extensively' about the precise wording of his statement" (Schubert 2007: 2). In fact, Kevin Rudd made good on his campaign promise by issuing an apology at the opening session of Australia's Parliament, on 13 February 2008. He said:

> the time for denial, the time for delay, has at last come to an end. Decency, human decency, universal human decency, demands that the nation now step forward to right an historical wrong. That is what we are doing in this place today (Cultural Survival Spring 2008: 7).

Tom Calma, Aboriginal and Torres Strait Islander Social Justice Commissioner, spoke after the Prime Minister, saying,

> Through one direct act, Parliament has acknowledged the existence and the impacts of the past policies and practices of forcibly removing Indigenous children from their families ... And by doing so, has paid respect to the Stolen Generations. For their suffering and their loss. For their resilience. And ultimately, for their dignity (Cultural Survival Spring 2008: 7).

It was not just Indigenous people who benefited from the change in government.

> Within 10 days of his Labor Party winning power ... Mr. Rudd had signed the Kyoto Protocol on reducing carbon emissions. Australia had been the only industrialized nation apart from the United States to resist doing so ... A week later, the new government scrapped another of Mr. Howard's ideological cornerstones: the ominously named 'Pacific Solution', under which hundreds of asylum-seekers were intercepted at sea, before they reached Australian waters, and shipped to remote Pacific nations to be processed (Marks 2008: 33).

Kathy Marks, Asia Pacific Correspondent for the British newspaper, *The Independent*, sees the apology as a "symbolic yet deeply significant gesture that many Aboriginal people described as the most important thing that had happened in their lives," and one that has meant a great deal to other Australians, sparking "a remarkable sense of national unity," with a Newspoll survey showing 69 percent voter support. This reflects a broadly based change in policy and approach. Among other appointments, Climate Change Minister Penny Wong is "Australia's first Asian-born minister and an openly gay woman" (Marks 2008: 33), and Rudd will hold "community Cabinet meetings" across Australia, in which members of the public will be invited to quiz ministers—a practice somewhat akin to the cross-country hearings linked to Canada's Royal Commissions.

I have firsthand experience of such processes. When I conducted the study for the Royal Commission on Electoral Reform, researchers had access to all of the transcripts from the town meeting-style hearings, and free rein to interview anyone we chose. The reams of transcripts I read included testimony from taxi drivers and teachers, hunters and artists,

Indigenous leaders and politicians, and many others. It was the most democratic process in which I have ever had the privilege to engage (Alia 1991a).

While the "historical significance of the apology to Aboriginal people cannot be underestimated," Marks cautions that the "national scandal of Aboriginal life expectancy, 17 years behind other Australians, remains" (Marks 2008: 33).

Australia has three major Indigenous publishers: *Land Rights News* published by the Central and Northern Land Councils, which, though mainly in English, is the only paper that has published in Aboriginal languages; *The Koori Mail*, owned by a group of Indigenous community organizations and based in Lismore, New South Wales; and *The National Indigenous Times (NIT)*, founded in 2002 and based in Canberra. "NIT publishes under the banner 'Building a bridge between Australia's black and white communities' while the *Koori Mail* proclaims itself as the 'voice of indigenous Australia'" (Robie 2005: 5).

Perhaps the earliest publication by an Aboriginal organization was *The Aboriginal*, also known as *Flinders Island Chronicle*, published in 1836. In 1938, *Abo Call: the voice of Aborigines* emerged in Australia—possibly the first 'advancement movement' newsletter—and an Indigenous supplement was published in the Torres Strait Observer (Meadows and Molnar 2002: 9). That same year, Torres Strait Islanders began experimenting with wireless transmitting. According to Ewen K. Patterson, this made them the "first natives in the world to have their own radio service," enabling the islands to "speak to each other and ... the outside world" (Meadows and Molnar 2002: 11).[1] Torres Strait Islanders are of Melanesian origin. The strait has more than 100 islands, 18 of which are inhabited by about 30,000 people from five main cultural groups (Shnukal 2001: 22). Since the 1950s, there has been a mix of publications in English and in various Indigenous languages. It was not until 1972 that the first Indigenous public radio program was launched in Adelaide, though there are "anecdotal reports" of Indigenous broadcasts that were heard in various locations across Australia, in the 1960s. Four Remote Commercial Television Services (RCTS) was licensed in the 1980s, though high transponder charges made them difficult to access. Imparja Television was created by CAAMA and several other organizations, named after the Arrernte word for "footprint" or "hunting tracks" (Meadows and Molnar 2002: 14). "Since their pioneering work with local 'pirate' television, the community at Yuendumu, along with three other communities in the Tanami Desert, developed other innovative uses for new technologies." The Tanami integrated "compressed

Figure 2.3. Uncle Bill Thaiday, pioneer of Indigenous broadcasting in Australia, at Woorabinda, central Queensland in 2005. On his right is Jim Remedio, Radio Manager of the Central Australian Aboriginal Media Association (CAAMA). Photograph by Michael Meadows.

Figure 2.4. Umeewarra Media radio station in Port Augusta, South Australia.

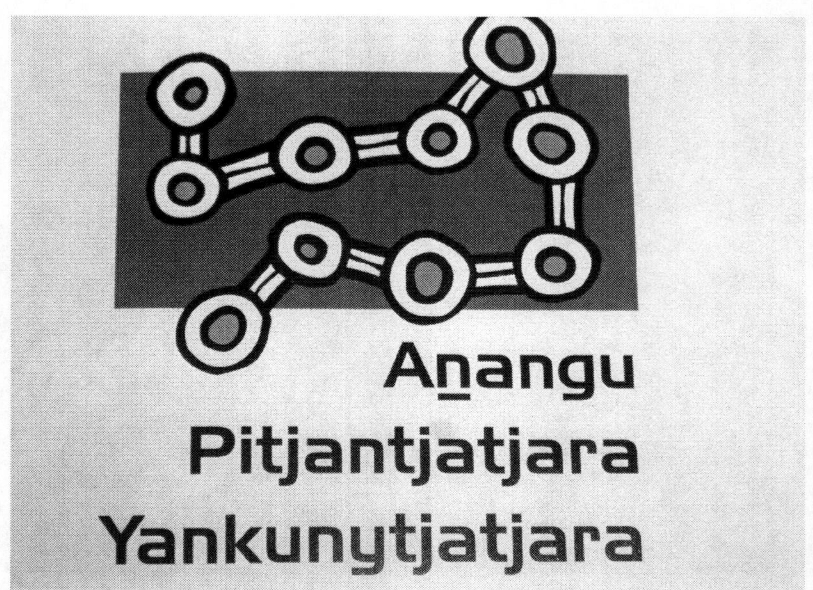

Figure 2.5. Logo for PY Media, the association that represents Anangu-Pitjantjatjara-Yankunytjatjara peoples of the central desert region, which spans the Northern Territory, Western Australia, and South Australian borders. Photograph by Michael Meadows.

Figure 2.6. Broadcast tent for Bumma Bippera Media, at the 2005 Laura Cultural Festival on Cape York, Australia. Photograph by Michael Meadows.

videoconferencing and satellite technologies based on three fundamental criteria: Aboriginal control; the need for a mixed package of media services (computer links, fax, telephone, local video production, broadcasting); and a wide application of technology to achieve cost effectiveness" (Meadows and Molnar 2002: 16). When the scheme became profitable, the profits were shared among the participating communities.

About 160 licensed community radio and television stations in remote areas "broadcast more than 1000 hrs of Indigenous content weekly as part of the Remote Indigenous Broadcasting Service (RIBS) [and a] further 50 or so community radio stations in regional or urban areas broadcast Indigenous content produced by local groups." There are 21 radio stations licensed as Indigenous-owned and run, three Indigenous low-cost narrowcast radio services and an Indigenous commercial station in Broome, Western Australia (Robie 2005: 6).

Australia's first exclusively Indigenous station began broadcasting in 1985 at Alice Springs. There are now stations at Brisbane and Townsville, an Indigenous community television station at Alice Springs, and the National Indigenous Media Association of Australia (NIMAA), with a membership of 136 community-broadcasting groups (National Indigenous Media Association of Australia 1998). The Western Australian Government introduced a video conferencing facility to spinifex people in a remote Indigenous community. "The Tjuntjuntjara community north of Rawlinna, on the Nullarbor, has been connected to the satellite technology in a bid to break down isolation barriers" (ABC June 2005). Traveling to meetings in Kalgoorlie takes about ten hours on a rough road; meetings in Perth or Canberra are even more challenging. While the motivation for providing the service may well be to save the government money and facilitate state and federal government meetings with members of the community, it also improves access for the people of Tjuntjuntjara to family, friends, and colleagues in other Indigenous communities.

In 2005, I met a few of Australia's many distinguished Indigenous academics, journalists, and artists and visited the National Indigenous Radio Service in Brisbane and Melbourne's Koorie Heritage Trust Cultural Centre. There, a powerful retelling of Australian history explores relations between the original inhabitants and immigrants through music, oral testimony, drawings, and photographs, and Genevieve Grieves' wonderful video installation, *Picturing the Old People*, offered a complex mix of powerful images, wit, and deeply moving moments. If, as Edward Said suggests, no one today is purely *one* thing, then all:

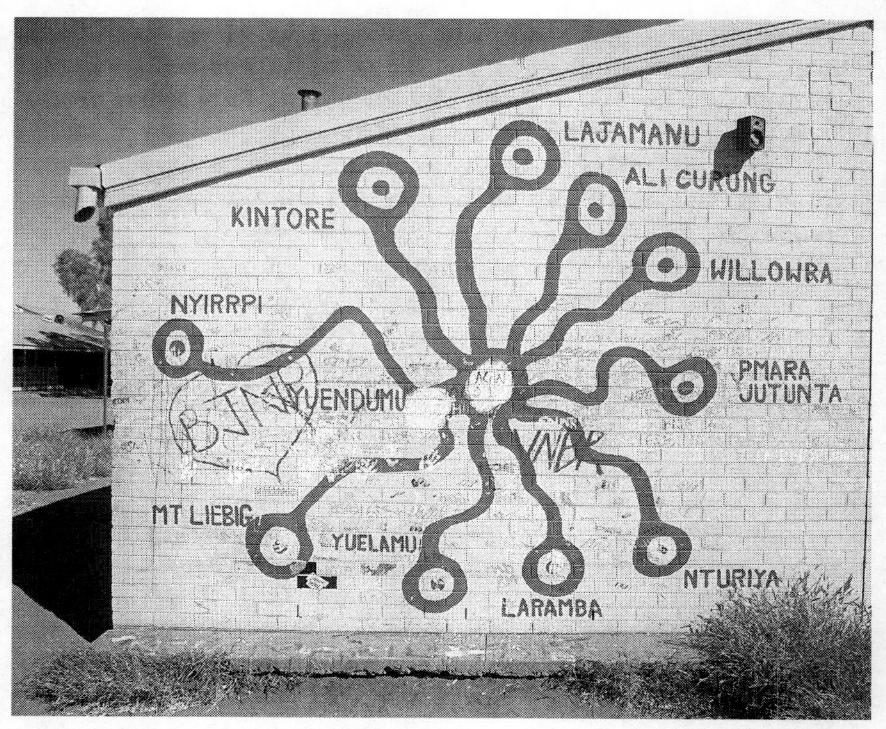

Figure 2.7. Mural at Yuendumu, 300 kilometres northwest of Alice Springs, showing the communities linked by PAW Radio. PAW stands for Pintubi-Anmatjerre-Warlpiri, the main language groups in the Tanami Desert of central Australia. Photograph by Michael Meadows.

> identities ... must be seen as dynamic. Aboriginality thus becomes a starting point as opposed to an ideologically constrained idea that defines people [in exclusive terms]. Survival is about making connections—cultural connections. And it is when Indigenous people have control of the means of production that a dynamic notion of Aboriginality ... can be identified (Meadows 2001: 206; Said 1993: 402).

The 150 or so Aboriginal and Torres Strait Islander groups working in community radio across Australia represent an enormously heterogeneous group—and this is reflected in their programming choices (Meadows 2001: 206). "The Indigenous cultures of Australia are considered to have the longest continuous history of any culture on earth." Today, there are more than 350 Indigenous languages (Bayagul 2007). Torres Strait Islanders are of Melanesian origins, and their ancestral homes are the many small islands located within the 200 kilometer strait between Australia's northernmost tip of Cape York and the southern coast of Papua New Guinea (PNG) (Beckett 1990: 1; Sharp 1992). Torres Strait became part of Australia in 1879. The movement for political independence began with the Maritime Strike of 1936 (Jull 1995; Sharp 1993: 181). In 1961, Islanders were granted full citizenship (Arthur 1998: 60; Beckett 1990; Kidd 1997; Jull 1997; Vlacci 1996), and in 2001, a new territory-style regional government was established within Australian borders, with administrative, funding, and legislative structures similar to those of Greenland Home Rule.

National Indigenous Media Services in Australia

The National Indigenous Radio Service (NIRS) "The Voice of Our People", is based in Fortitude Valley, Queensland. Based in Brisbane as a hub station linked to its primary network via a series of digital ISDN lines. Delivered via a national satellite footprint, it can be received anywhere in the country, and plans are under way for a national radio news service. Full membership is open to those Indigenous community-controlled organizations which are actively involved in community broadcasting, radio, film, video, television and print media. Associate membership is open to all Indigenous individuals who are actively working in media. Produced in 2005, NIRS' Draft Indigenous Media Code of Ethics includes provisions for respecting " all confidences received in the course of their calling, especially secret sacred information" and for declaring conflicts of interest where family and clan loyalties may be concerned. It also calls for respect for "private grief and personal privacy," for educating the wider community with regard to Indigenous media and issues, and for observing "all tribal, geographical and sacred rights and areas" (NIMAA 2005).

In 2007, the Powerhouse Museum in Sydney featured an exhibition by Bayagul: Contemporary Indigenous Communication, an association whose name means 'speaking up' in the Eora language. Here, 'communication' is defined in its broadest sense, as represented by the Marrugeku Mimi dancers, the architectural design of the Merrima group, and Imparja TV, Australia's first Indigenous television station (Bayagul 2007).

> *Bayagul* reveals aspects of Indigenous Australian identities as they are expressed through today's technologies and industries ... Aboriginal and Torres Strait Island peoples are speaking up for themselves in film, dance, music, architecture, tourism, media, fashion and much more! ... *Bayagul* also speaks through the work of outstanding contemporary Indigenous artists and performers such as Jimmy Little, Bronwyn Bancroft, Tracey

> Moffatt, Rachael Perkins, Deborah Mailman, Justine Saunders, Bangarra Dance Theatre and Mervyn Bishop (Bayagul 2007).

In 2004, the Australian Government produced an eye-catching leaflet titled, "*Welcome to country: Respecting Indigenous culture for travellers in Australia.*" It calls on visitors to behave with appropriate consideration and respect with regard to the Indigenous peoples of Australia and their "many different Indigenous language groups and cultures." It notes that privacy is paramount, and some communities will not welcome uninvited guests, or wish to have the land and waters disturbed by seekers of "mementos" (Aboriginal Tourism Australia 2004, unpaginated). A couple of years after the government campaign, a company called Travelmood produced a truly offensive advertising campaign. It features gorgeous, full-color photographs of Uluru/Ayers Rock in all its sunlit glory. In the midst of sunlit brush, front and center, we see a table with sparkling white tablecloth and napkins, a champagne cooler and glasses, and two upholstered chairs. A blond-haired woman is serving the imaginary diners in the empty chairs. Beneath the photograph, rows of dots hint at Indigenous art. The caption tells us to "EXPERIENCE AUSTRALIA'S CULTURAL HEART." What culture? What heart? (Travelmood 2007).

One of the most powerful cross-cultural communication tools I have seen is the map of *Aboriginal Australia* compiled by David Horton (2000) for the Australian Institute of Aboriginal and Torres Strait Islander Studies. For the non-Indigenous person, it requires a massive mind-shift that entirely changes one's understanding of the place called "Australia." Big blocks of color identify cultural and linguistic groups. The eye is drawn to a large territory labeled "Yuggera" and the neighboring "Waka Waka," before moving to the smaller space and gray print identifying the colonial city of Brisbane that sits in the midst of Yuggera land. The country is massive, and the blocks of color identifying its many-cultured Indigenous population dwarf the settlements of newcomers.

Aotearoa/New Zealand

In 2005, the Māori filmmaker, Barry Barclay—best known for his landmark feature film, *Ngati*—reflected on the challenges and rewards of intercultural exchanges and transcultural communication:

> I believe—passionately, as it happens—that it is possible to share with other peoples our own works and to be given opportunities to enjoy their works in return. We cannot afford to be light-headed about it, though, for experience teaches us that, when moving as artists into the Indigenous

> world, we may unwittingly be the occasion of significant hurt. With a little respect and understanding, we can avoid that (Barclay 2005).[2]

Following his death in 2008, he was honored at the closing ceremony of the inaugural World Indigenous Television Broadcasting Conference, March 28, in Auckland, as a "pioneer of Maori television and film":

> Barclay (1944–2008; Ngāti Apa), ... was posthumously awarded the inaugural Te Puni Kōkiri Lifetime Achievement Award for Indigenous Television Broadcasting, Te Rerenga Tahi ... launched as part of WITBC '08—a three-day gathering of indigenous television leaders from throughout the world ... hosted by New Zealand's national indigenous broadcaster, Māori Television ... Barclay was the influential director of Television New Zealand's first major documentary series on Māori life and culture, *Tangata Whenua* (1972) ... In the late 1970s and early 1980s, he worked ... on projects in Sri Lanka, London, Paris and Amsterdam before returning to New Zealand to write and direct *The Neglected Miracle*—a feature-length political documentary on the ownership of plant genetic resources, shot over two years in eight countries. In 1987 [he] became the first Māori to direct a dramatic feature, *Ngāti*, which won Best Film at the Taormina festival. In 1991, he wrote and directed the feature *Te Rua* ... about a group of Māori who set off for Berlin to claim back tribal carvings held in a museum ... His most recent film was *The Feathers of Peace*, a feature drama-documentary ... on the Moriori people of Rekohu (the Chatham Islands) ... The award was presented during tonight's one-hour live-to-air launch ... of Māori Television's new 100 per cent Māori language channel, Te Reo (Horan 2008).

Māori arrived in what is now Aotearoa (the Māori name for New Zealand) around 1200 AD from Polynesia. Today, 85 percent live in urban areas (New Zealand 2004: 7). The Dutch were the first Europeans to arrive, starting with Abel Tasman in 1642, followed by the English, led by Captain James Cook in 1769 (New Zealand 2004: 12–13). After a protracted struggle, an agency, Te Māngai Pāho, was established in 1993 to fund the promotion of Māori language and culture, but it took more than a decade for Māori television to arrive. "The unexpectedly high core viewership per month (of which at least 60% is estimated to be Pākehā [non-Māori]) suggests that ... tikanga and te reo Māori have a pivotal function to play in not only te ao Māori, but also the nation as a whole." Smith and Abel note, "The increasingly central role that audiovisual culture plays in the negotiation of community relationships, social power and cultural survival" and locate Māori Television within "a global indigenous media movement." A second, more iwi-based television channel is planned (Smith and Abel 2007).

Māori Television Service (Te Araratuku Whakaata Irirangi Māori) Act 2003 states that "the principal function of Māori Television is to promote te reo Māori me ngā tikanga Māori [Māori language and customs] through the provision of a high quality, cost-effective Māori television service, in both Māori and English, that informs, educates and entertains a broad viewing audience, and, in doing so, enriches New Zealand's society, culture and heritage" (Te Araratuku Whakaata Irirangi Māori 2003). Following *iwi* radio, spurred by the 1986 Waitangi Tribunal claim, Māori demanded control and ownership of their own TV channel (Mita 1996: 43). In March 2004, after years of lobbying, the Māori Television Service launched a dedicated Māori television channel. At the same time, Māori language, culture, and primetime programming are expanding on the mainstream networks (New Zealand 2004: 14; Alia and Bull 2005). Between the early 1950s and late 1960s, 80 percent of the Māori population migrated to cities, especially Auckland and Wellington, mostly in search of jobs (Walker 1975). The effects are captured brilliantly in Barry Barclay's film, *Ngati* (Barclay 1996).

Once urbanized, more Māori started attending university. At the same time, through television and other media, they were exposed to student political activism, the anti-war movement, the US civil rights and African American movements, women's liberation, and the gay and lesbian rights movements (Smith and Smith 1990; Poata-Smith 1996). Drawing on those developments, Māori intellectuals became increasingly politicized and—as in Greenland, Nunavut, and Sámi countries—an explosion of journalistic and political activism accompanied the 'Māori Renaissance' (Spoonley 1990; Walker 1992).

A BBC broadcast in 2004 reported:

> About 10,000 mainly Maori protestors have marched through the New Zealand capital against plans to nationalize the country's beaches and seabed… At Wednesday's march, fifty Maori warriors led by organizer and academic Pita Sharples performed a *haka war dance* (my italics).

Pākehā audiences may well have thought they were learning both about New Zealand news and Māori culture. The problem is that a haka is not a dance, and this particular haka was performed to express dissent, not to provoke war. Recent films provide indigenous perspectives tinted—or tainted—with colonial accounts. Lee Tamahori's 1994 film version of Alan Duff's novel, *Once Were Warriors*, presents a shocking portrait of urban Māori family life that relies heavily on the stereotypical view of

Māori violence derived from a warrior past. Some have suggested that, rather than inhibiting violence, this portrayal may even have encouraged some Māori to commit crimes, in the belief that they are following their warrior heritage (Alia and Bull 2005).

In some ways, the appearance of *Once Were Warriors* was a defining moment in New Zealand history. The film was watched by at least a third of New Zealanders and a substantial international audience (Television New Zealand 1999). The story takes place in a poor urbanized area of south Auckland. While the focus is on culture, questions of class lurk just below the surface. The wider implications of class differences are never addressed. By representing Māori as hell bent on their own destruction, *Once Were Warriors* plays directly into the hands of the mass media, which have already constructed minority peoples as criminals. The Māori 'tradition' invented by Pākehā is motivated by an assimilationist agenda. The Māori tradition that Māori invent counters Pākehā values such as individualism and materialism.

The Māori "language nest" movement, Kohanga Reo, has had worldwide influence on Indigenous and other minority language immersion programs, such as the Aha Punana Leo progam in Hawai'i (Arno 2008). In Aotearoa, it involves the entire *whanau* (extended family) with children, parents, and grandparents sharing in the learning experience, using Māori teaching-learning methods and curriculum developed by Māori. It originated in 1980 with Māori elders, and opened formally a year later. Today there are about 600 centers. The guiding principle is: "*Ko te reo te mauri o te mana Maori*—the language is the life principle of Maori dignity" (New Zealand 2004: 49–50; Alia and Bull 2005).

Indigenous broadcasting is more recently arrived in New Zealand, where 21 Māori radio stations are linked to Ruia Mai—the national Māori radio service founded in 1990 (Alia and Bull 2005). All are bilingual, and they are required to broadcast minimum of 30 percent in te reo Māori to qualify for state funding. There are no surviving *iwi* newspapers. There is one bilingual Māori paper, *Te Māori News*. More than 95 percent of New Zealand households have telephones; more than half have mobile phone and Internet access. In addition to TVNZ, the government supports *Te Mangai Paho* (Māori Broadcasting), the fledgling Māori television channel that broadcasts entirely in *te reo Māori* (Māori language) (New Zealand 2004: 58–59). Linda Tuhiwai Smith reflects on the relationship between Māori media and mainstream media, language, and culture:

> In the township that I come from … [television] imports American culture and educates the tastes of the young for labelled clothes and African

> American rap. In the town is the local tribal radio station which beams out the news and local music in the tribal dialect of both English and Maori languages. Local *iwi* politics are discussed regularly along with the finer details of the most recent rugby games (Tuhiwai Smith 1999: 96).

The Pacific Islands

The first Pacific Island radio service was launched in 1933 in Papua New Guinea. Two years later, radio began in Fiji. "These stations were run by government information offices and provided very limited services to local communities." Radio did not reach out to other Pacific countries until the 1960s, starting in Niue in 1967 and in Nauru in 1968. As Molnar and Meadows emphasize, "The history of Indigenous broadcasting in the Pacific, as it is in Australia, is very recent" (Molnar and Meadows 2001: 77).

Indigenous media have experienced continuous under funding and a general lack of government support. "Across the board in Fiji and Papua New Guinea—and elsewhere in the Pacific—salary structures are pretty appalling," as a Papua New Guinean journalist described it. This view is shared by more than two-thirds of the 57 reporters, media executives, and policymakers who were interviewed. Despite the observation of this general pattern extending throughout the Pacific region, there are significant differences, which the researchers cannot entirely explain. For example, although journalists in Papua New Guinea are generally better educated and have a much wider range of experience than those in Fiji, "they are far more poorly paid" (Robie 2006: 17).

Robie questions the extent to which media can remain independent, in light of the influence of the rampant "'envelope journalism' inducements by unscrupulous politicians" that he has observed in Indonesia, the Philippines, Papua New Guinea, and the Solomon Islands. A Port Moresby journalist told Robie: "If you're working for an absolutely meagre wage … your opinions are going to be able to be changed or swayed with a bit of influence … a bit of cash or some sort of incentive … I think it is rife" (Robie 2005: 18). Despite this frustrating and debilitating situation—or perhaps because of it—"almost three out of every four PNG journalists regarded 'communicating knowledge in the community' as a crucial factor in taking up a media career" and many of the journalists cited the objective of "exposing abuses of power and corruption" as a reason to become journalists (Robie 2005: 18–20).

Sápmi

Observing that "failure" can lead to new kinds of success, Heininen comments on Indigenous responses to national borders, using the example of Sámi:

> [The] Sápmi homeland of the Saami people is divided by the national borders of four different unified states. In 1980–81, the Alta movement against the harnessing of the Alta River in northern Norway mobilized Saami across the national borders to reassert their identity as an indigenous people and to strengthen their demands for self-determination ... Although this radical transnational movement lost its fight over the dam, it spawned a national awakening, especially among young Saami and Saami artists (Heininen 2004: 209).

"People began to arrive in Fennoscandia around 11,000—10,000 BC ... cultural zones developed very early on and were characteristic of later periods as well." Today, there are 80,000–100,000 Sámi in Norway, Sweden, Finland, and Russia's Kola Peninsula. The numbers are approximate because decades of discrimination and racism have led some Sámi to refrain from identifying as Sámi, making it hard for census takers and researchers to obtain accurate data. The greatest numbers are in Norway—approximately 40,000–60,000, with 15,000–25,000 in Sweden, 7,000–10,000 in Finland, and 2,000–4,000 in Russia (Aikio-Puoskari 1998: 47). State borders did not begin to emerge until the thirteenth and fourteenth centuries. Progress in Sámi-state relations has not been linear. For example, Veli-Pekka Lehtola points out that the "Swedish Crown recognized the siida [Sámi village system] boundaries and ownership rights until the 1700s," an observation that is particularly interesting in light of more recent history, when Sámi have had relatively less power in Sweden than in other areas of Sápmi (Lehtola 2002: 23).

Cultural intervention was particularly strong in Norway, where the "Norwegianization policy" was based on the premise expressed by Johan Sverdrup, that the "only way to save the 'Lapps' is for them to merge into the Norwegian People" (Lehtola 2002: 45). Lehtola calls this "*active* colonialism" through which "Sámi were clearly to be assimilated into Norwegian society and they were to obliterate the Sámi language." Sweden developed what it called "the segregation and isolation policy," which, on the surface, seemed more moderate than Norway's but was nonetheless a policy of "implied assimilation" that depended on the acknowledgement of the Sámi as a distinct group (Lehtola 2002: 45). "The ideas of social Darwinism reached Finland later than elsewhere in the Nordic countries [in] the 1920s and 1930s." Lehtola attributes this to Finland's independence in 1917 and its need "to seek its own identity as a state" (Lehtola 2002: 46).

Sámi have long had a propensity for organizing into associations, local and regional organizations. According to Lehtola, 'the peak of Sámi social activism' was a meeting held in Trondheim, Norway on 6 February 1917, "the first attempt to gather Sámi representatives across borders" (Lehtola: 46–48). From the early days, Sámi political organization has been accompanied by the emergence of Sámi media, starting with early newspapers and continuing to recent developments in radio, television, and new media. The first Sámi political newspaper, *Sagai Muittalaegje*, began publishing in 1906 (Lehtola 2002: 48–49).

In recent years, a strong sense of Sámi identity and location in Sápmi, "Land of the Sámi"—the Sámi homeland—has increasingly transcended the boundaries artificially imposed by nation states. As Tove Anti, staff officer with the Norwegian Sámi Parliament, put it: "We never say, 'We are going to Finland' ... We say, 'We are going to the Finnish side of the border.' We also don't call this the Norwegian Sami Parliament; we call it the Parliament of the Samis in Norway. It's important because words have a lot of power" (Anti in Hoge 2001: 2). Contrary to stereotype, only about ten percent of Sámi are reindeer herders; others fish and are engaged in other occupations.

The Áltá Conflict was a major turning point in Sámi political and cultural life. It was in some ways a last straw reaction to environmental, cultural, and political disruption caused by the building of large dams, starting in the 1930s, throughout Sápmi, in Norway, Sweden, Finland, and the Kola region. Sámi protests were most sharply focused between 1968 and 1982, on attempts to protect the Áltá-Guovdageaidnu River system from government plans to build a massive dam and power plant. Although the dam and plant were built, with serious environmental consequences for the entire region, the Áltá Conflict and the resulting movement have had lasting effects on Sámi political life, with continuing collaboration between artists, political, and social leaders, and what is known as the 'cultural awakening' that parallels the Maori Renaissance and what I have called the Inuit Renaissance. A national symbol emerged—the Sámi flag—first flown during the Áltá Conflict and made official at the Sámi Conference in 1986, with its colors of yellow, red, and blue in a circle "signifying unity" (Lehtola 2002: 56). In Lehtola's view,

> The Áltá Conflict was remarkable politically because it caused Norway to change its official policy toward Sámi ... Within one decade of the Áltá Conflict, Sámi politics accomplished a large part of the reforms that the

> Sámi had demanded during that struggle. [They] were recognized as an aboriginal people under the amended Norwegian Constitution in 1988; obtained their own representative body, the Sámediggi [Parliament] in 1989; received their own political advisory authority to the government [and] secured the Sámi language act in 1990 (Lehtola 2002: 72–73).

Along with this, Sámi have suffered "environmental catastrophes which they had no possible way of influencing," including radioactive rain and other long-term effects of the 1986 Chernobyl disaster in the Ukraine.

Helsingin Sanomat (*HS*), the major Finnish newspaper, "acts as a semi-official voice in the public sphere." Pietikäinen conducted a study of 51 news stories about Sámi, published in HS from 1985–1993—a period that "captures a phase of transition in the construction of Sami identity; both the strengthening of Sami resistance politics and ... increasing contestations of their rights and indigenous status. Among the topics that were virtually non-reported were relations between the Sami and the ethnic majority living in the Sápmi—at times a tense and heated situation. Also, background information on Sami culture was seldom given..." (Pietikäinen 2003: 591–593).

Along with Maori, Canadian First Nations and Inuit and Indigenous Australians. Sámi were among the founders of the World Council of Indigenous Peoples (WCIP) in 1975.

Japan

In 1593, the warlord Toyotomi Hideyoshi gave the Matsumae clan rights to the island of Ezo, now Hokkaido, the ancestral Ainu homeland. The Hokkaido Colonial Office was set up in 1869 and in 1899, the Meiji government enacted the Hokkaido Former Aborigines Protection Law. Much as 'reserves' in Canada and 'reservations' in the United States were created to herd Indigenous people together and free up the land for colonial use, in Japan, "certain tracts of land were designated for Ainu people, their common property placed under the control of the governor of Hokkaido, and education guidelines outlined by the state" (Zakoda 2006). There are about 15,000 Ainu in Japan, though it is said that many more may choose not to identify themselves as Ainu, and change their names to avert discrimination, which Ainu have experienced for centuries (Temman 2006).

A 1990 exhibition in Daito-city Osaka that toured to other parts of Japan and to New York, San Francisco, and Los Angeles expressed hope along with anger and frustration. "Against the plunder and suppression of Japanese, [segregation], and infringement on human rights, [we stand] up bravely to say 'Yes, in spite of everything, We Ainus are alive today' … we indigenous peoples [have] no 'borders'" (Yamamoto 1993: 5). The exhibition included Mieko Chikkap's poem, "We may hear the pulsation of the 21st century beating for our future":

> Break down "apartheid"!
> Risking our lives,
> We'll fight!
> The day of victory will surely come.
> Break down "apartheid"!
> …
> The *Nibuani* valley,
> The sacred place [of] *Ainu . moshiri,*
> Is going to be destroyed.
> A dam is now under construction there!
> Do they say we have no racial sovereignty?
> Don't let the erroneous history repeat
> Itself any more!
> Stop mining coal and uranium ore!
> …
> Look!
> Look at the earth with open eyes!
> She has been sobbing…
> …
> let's sing for tomorrow…
> …
> I feel bright sunlight on my eyelids.
> I hear the birds singing.
> A tear drop is on my cheek.
> Was it a dream…
> (Chikkap 1993: 13–15)

Shigeru Kayano, who died in 1979, whom some called the "Ainu Mandela," was an ethnographer, historian, and author of more than a hundred books, and "symbolic leader of the Ainu, from whose dialect he drew a dictionary," on whose behalf he won a seat in Parliament in 1994. He drafted the Law for the Promotion of Ainu Culture, adopted by parliament in 1997 (Temman 2006). His strongest legacy is likely to be the fostering of Ainu media—most directly, the creation of the Ainu radio station in his home community of Nibutani. Nibutani has become a global gathering place for Indigenous peoples. In 1993 and 2005, it hosted the Nibutani Forum to discuss how Indigenous peoples can work together (Zakoda 2006). In the summer of 2008, following the formal apologies in Australia and Canada, the government of Japan acknowledged Ainu rights.

"After centuries of discrimination and forced assimilation, Japan's Ainu people have finally been recognized by Japan's government." (Cultural Survival July–August 2008). However, the Japanese government is still far from fulfilling its responsibilities under the UN Declaration on the Rights of Indigenous Peoples (Indigenous People's Summit in Ainu Mosir 2008), and media scholar Gabriele Hadl cautions against excessive optimism. Ainu radio broadcasts only two hours a week, and with an access range of only a few hundred meters, requires no license. Hadl does not think they are likely to obtain a "regular license" in the foreseeable future (Hadl 2008). Shigeru Kayano's son, Shiro Kayano carries the torch. He chaired the organizing committee for the Indigenous People's Summit in Hokkaido in advance of the G8 ministerial and has continued FM Pipaushi, and expanded the network. From a recent study group meeting in Japan, a researcher reports: "He just visited Indigenous radio stations in Bolivia, Mexico and Columbia and returned inspired" (Hadl 2008)—an example of the New Media Nation in action. Encouraged by the experiences of Indigenous people in Canada, Greenland, Australia, Russia, Aotearoa-New Zealand, Sápmi, and many other places, Ainu are experiencing the beginnings of a cultural rebirth. Among the leaders are members of a group called the Ainu Rebels, founded in 2006 by Mina Sakai, her brother, Atsushi, and a group of friends. "Mina Sakai has succeeded in an unconventional way in boosting public interest in the indigenous people ... She performs traditional Ainu dances and music mixed with rock and hip-hop." The group moved to Tokyo from Hokkaido, the ancestral Ainu home. "Like many other Ainu activists, Sakai first used political means to advocate Ainu rights, including taking part in the U.N. working group on indigenous populations...Shizue Ukaji, a well-known Ainu activist and artist, ... encouraged her" to combine politics and art. Sakai told a journalist,

"I believe that entertainment can help us easily cross any boundaries between people, and I want people to first learn about the positive aspects of the Ainu, before [they learn] discrimination and prejudice"... he Ainu Rebels have succeeded in drawing attention to the Ainu with their...performances while giving audiences an opportunity to think about the social issues.

In 1998, she visited Canada through a program organized by an Ainu organization, and met Indigenous people in British Columbia and Alberta. She said,

> "I could tell how proud they were of themselves, their ethnicity and their origin ... I realized that our people could also live that way."
>
> [She] attends an Ainu language course [because] The Ainu Rebels sing in their ancient language, but none can speak it because their ancestors were not allowed to pass it on under the past assimilation policy of the government (Yasumoto 2007).

Kalaallit Nunaat/Greenland

Inuit have lived on the coastal areas of Greenland (Kalaallit Nunaat) for about 5,000 years. In 1979, Greenland's Home Rule Act took effect, and the *Landsting* (Greenland parliament) and *Landsstyre* (Greenland government) sat for the first time (Berthelsen 1995: 15). Under Home Rule, Greenland became a special cultural community within the Kingdom of Denmark.

As Aboriginal broadcasting was emerging in the Canadian North, Greenland was also enjoying its own communications revolution, which like that of Canada featured a Kalaallit (Greenlandic Inuit) cultural renaissance. In 1978 Kalaallit Nunaata Radioa (KNR)—Greenland's public national broadcaster—was made an associate member of the European Broadcasting Union, thus legitimizing Aboriginal Greenlandic radio and television in the eyes of European broadcasters. Today, KNR has its headquarters in Nuuk (the capital) and regional offices in Qaqortoq and Ilulissat. It broadcasts a daily radio newscast from a station in Copenhagen, for Greenlanders living in Denmark. KNR's television service is linked to Danmarks Radio (DR), Danish public radio, which distributes its newscasts throughout Greenland via satellite. Greenland also has a handful of privately owned radio and television stations. KNR Radio and KNR Television provide an increasing proportion of their programming in *Kalaallisut* (the Greenlandic language). In June 2009, following a referendum, a new era of Self Rule began. One of its most notable changes was the establishment of *Kalaallisut* as Greenland's official language. While Self Rule does not constitute complete separation from Denmark, Self Rule, and the shifting linguistic and cultural landscape, move Greenland closer to that possibility. Danish people sent and controlled the early broadcasts, but *Kalaallit* (Greenlanders) were soon involved, and by the 1950s, programs were produced in Greenlandic and Danish.

Broadcasting began in Greenland in 1925 with the transmission of the first telegram. In 1980, broadcasting was one of the first institutions to come under the control of Greenland Home Rule jurisdiction. KNR emerged in the midst of a cultural revival that included much of the music broadcast on its radio outlet. Greenland's singer-songwriters were among the first to write and record in their own language, and have influenced indigenous musicians worldwide. Of the approximately 5,400 hours of broadcasting annually, some 2,200 hours are devoted to music, supported by Greenland's strong and rapidly expanding music recording industry. In addition to music broadcasts, KNR's radio and television outlets feature a wide range of regional, national, and international entertainment, news, public affairs, cultural and youth programming. As Hans Lynge put it, radio broadcasts soon became "as necessary a part of life as food" (Alia 1999: 94).

Latin America

Indigenous resistance movements in Latin America date back at least to eighteenth-century uprisings in Peru and Bolivia, and continued in the nineteenth and early twentieth centuries, participating in the Mexican Revolution and in several other wars for independence. After decades of suppression, Indigenous peoples are experiencing a rebirth of social, cultural, and political activism, supported by emerging media outlets, experiments, and organizations. Their work is supported by what Juan Francisco Salazar terms "a cultural appropriation of audiovisual electronic and digital media" (Salazar 2003). Salazar is both a media anthropologist and an activist who is one of the coordinators of the international OURMedia network and an associate of the Coordinadora Latinoamericana de Comunicacion de los Pueblos Indigenas (CLACPI).

Latin American Indigenous people are rapidly developing media networks, film, video, radio, and online media, and consider these media projects a catalyst, not a substitute, for traditional organizing (Delgado 2003). Language, media and communication rights are on the rise. In Ecuador, the official languages are Kichwa, Shuar, and Spanish. Colombia was one of the first countries to change its constitution to acknowledge Indigenous rights, including those of communication. The constitutions of Venezuela, Colombia, Ecuador and Bolivia, guarantee Indigenous peoples the right to their own communications media. An Indigenous video movement began in 1996, when the Latin American Indigenous Film and Video Collective (CLACPI) held its first Indigenous Film and Video Festival in Bolivia. Since then, CLACPI, a network of Indigenous video practitioners and activist, has screened more than four hundred videos at festivals in Guatemala (1999), Chile (2004), Mexico (2006), and again in Bolivia (2008).

Chile passed its new Audiovisual Law in 2004. It makes no reference to Indigenous media/communication. But despite government efforts to suppress Indigenous (particularly Mapuche) activists, Mapuche online media, community radio, television and video-making have survived, and thrived (Salazar 2003). Mapuche have developed an extensive, online digital network that is accessed throughout Chile, and internationally and includes a digital newspaper, *Azkintuwe*, the online *MapuExpress*, and *Mapuche International Link*. Latin America's Indigenous media movement has shown that mediamaking "is a critical form of making politics" (Salazar 2003; Salazar 2008).

The United States

US policies have vacillated wildly over the years, ranging from the assimilationism that is often called "melting pot" politics, to various forms of active and more passive genocide, such as the decimation of Native American peoples that occurred during the forced relocations from various parts of the country to Oklahoma, via the "Trail of Tears." The American Indian Movement that emerged in the 1960s was both a political turning point and part of a cultural renaissance. In the decades since, policies have shifted to allow for major changes to museum and other institutional policies, with repatriation of human remains and cultural artifacts a major theme.

In general, literature—non-fiction, fiction, and poetry—has thrived, along with music and the visual arts, though in general, the arts labeled "handicrafts" have predominated and the works of other Indigenous artists have tended to be more anthropologized and less widely known and exhibited than those in Canada. The earliest prominent writer may be Black Elk (Hehaka Sapa), an Oglala Sioux born in 1863. Among those on a long list of outstanding writers, many of whom figure in what Kenneth Lincoln called the "Native American Renaissance" (Lincoln 1983) are Louise Erdrich, Anishnaabe (Ojibway or Chippewa) Nation and German-American; M. Scott Momaday (Kiowa), who won a Pulitzer Prize in 1968 for his novel, *House Made of Dawn*; Joy Harjo, Creek Nation poet and musician known for her delightfully named band, Poetic Justice; Paula Gunn Allen, who is of Laguna, Sioux, Scottish, and Lebanese-American descent; Leslie Marmon Silko (Laguna Pueblo tribe); Thomas King (Cherokee, Greek, and German Canadian and American); Velma Wallis (Athabaskan Alaskan); Sherman Alexie (Coeur d'Alene); Dian Million (Alaska Native); and Gloria Bird—whose edited collection with Joy Harjo of works by Indigenous women, *Reinventing the Enemy's Language*, itself expresses the work of their generation. Other writers cross the US-Canadian border, among them Lee Maracle, Drew Hayden Taylor, and numerous others. As in other parts of the world, Indigenous artists in the US often cross various boundaries. For example, the actor, singer, songwriter, and AIM (American Indian Movement) activist, Floyd Red Crow Westerman (who died in 2007), and the actor and journalist, Gary Farmer.

Some of the US media developments are reported in the chapter on Canada, as the border crossing often makes it hard to separate people on the two "sides" of the US-Canadian border—and indeed, this is one of the main points that Indigenous people in North America and elsewhere keep emphasizing.

Some experiments and associations are unable to sustain themselves on one side or the other, and are transformed. One example was the attempt to establish an organization of Indigenous journalists in Canada. The Native Journalists' Association (NJA) survived for only a couple of years. Unable to sustain itself alone, it disbanded and many of its members joined the US-based Native American Journalists' Association (NAJA). While on the surface, it was a response to "failure," the cross-North America link made a point that many Indigenous people have been making all along: the cultural and political divisions between Indigenous people in Canada and the United States have been artificially created and manipulated by state governments. There is a legitimate political point to be made by linking the journalists in both countries (and in turn, with Indigenous journalists internationally).

However valuable that may be, the reality is also that distance and expense mean that Alaska Natives and many Aboriginal Canadians have had a hard time accessing the invaluable networking, information, and skills sessions featured at NAJA conferences. In 1998, the NAJA conference included an energetic session on the relationship between Aboriginal publications and First Nations governments. The year before, Alaska's landmark Indigenous newspaper, *Tundra Times* (nominated for a Pulitzer Prize under the leadership of its founder, Howard Rock), ceased publication. Lael Morgan attributes this to a drop in credibility, circulation, and advertising, a focus on "public relations handouts" and a policy of allowing "only positive news about native people" (Morgan 1998: 33–34). Morgan's assumptions cannot be read uncritically. Her figure of 80 percent public relations content (reported in 1998) was based on a 1989 study, and there was no evidence correlating "positive" news with poor circulation. In fact, Morgan's comparison of the *Times'* "public relations" coverage with the non-Native, Pulitzer-winning *An-*

chorage Daily News' studies of the "native suicide rate and alcohol abuse" hit a sore spot. In private discussions, tribal editors, writers, and publishers complain that non-Native journalists win prizes for "mostly negative," inadequately researched portrayals of Native American people, issues, and communities (Steffens and Alia 1997–98). The first Indigenous broadcasts in North America were heard on Alaskan radio in the 1930s, but despite Alaska's nearly 30-year head start, Canada is the world leader in Indigenous broadcasting. The United States has moved all too slowly to support Indigenous media. In late 2006, the population of the United States was about 298,500,000, and that of Canada, a mere 38,000,000. Yet Canada had several hundred Indigenous radio stations, as compared to only about 30 in the United States. Canada also has eleven regional networks, a national radio network, at least six television production outlets, and the Aboriginal Peoples Television Network. While still not correcting the Canada-US imbalance, Native American radio took a big leap forward between 2007 and 2009. A campaign to increase Native American media ownership by the advocacy and broadcasting organization, Native Public Media, led to fifty-one new license applications by thirty-eight tribes and organizations. By 2009, twenty-nine licenses had been granted, bringing the number from 33 in 2007 to 62 (Taylor 2009).

There was also a small step towards introducing Native American television, with the launch in 2008 of the first tribal cable network by the Tulalip Tribe in western Washington state. The cable channel provides programming to the local cable station, KANU-TV 99, and to the tribe's Internet and cable provider, Tulalip Broadband, and is available internationally as a live webstream. There are no commercial Native-American-owned television stations (Taylor 2009).

As in Australia, Sápmi, and many other parts of the Indigenous world, in North America, radio rules. Even in relatively wealthy "first world" countries such as the United States, it is a mistake to assume universality of access to a range of media and "new media." As Levo-Henriksson points out in her study of Hopi media and identity, many Native American communities have poor, and sometimes no access to relevant radio, television or the Internet. Before 2000, the Hopi Reservation "was not able to take full advantage of telecommunications. The electronic infrastructure was not there," and as of 2007, most Hopis "still have no access to e-mail or other online services" (Levo-Henriksson 2007: 107). This is not due just to poverty and the resultant scarcity of computers in Hopi homes, but to the fact that the United States does not have a culture of shared community access comparable to that found in the Canadian North and other Canadian regions. Neither the federal or state governments nor the tribal governments seem to prioritize communication infrastructures or community access.

When I first went to what is now Nunavut in the early 1980s, even the smallest Inuit communities were providing equipment, access, and expertise, and soon were developing websites and exploring various options for using the new technologies. Inuktitut fonts were available by the mid 1980s and computers were provided to schools, adult education centers, and community centers. Teleconferencing was emerging as a major tool for conducting inter- and intra-community meetings, legal, medical, and social services and government business in a region with challenging weather, distances, terrain, and prohibitive transportation and communication costs. Radio, television, and Internet resources were already in place and in the decades since, have continued to expand and develop.

By contrast, relevant broadcasting did not come to the Hopi Reservation in northeastern Arizona until the cusp of the new millennium. As late as December 2000, Levo-Henriksson attended the launch of KUYI-FM, the Hopi radio station on First Mesa. Created by The Hopi Radio Project, its aim is "to bring a Hopi-owned, Hopi-run FM public radio station to the reservation, in part to help preserve the Hopi language and culture" (Levo-Henriksson 2007: 103). That said, Levo-Henriksson did find computers in some Hopi schools, and concomitant Internet access, in the spring of 1996. Continuing scarcity, however, means that most Hopi access the Internet from the city of Flagstaff or other outside locations (Levo-Henriksson 2007: 148). Levo-Henriksson's study raises important questions about pan-indigeneity, strategic essentialism, sociopolitical priorities, and linguistic and cultural integrity. Whether the Hopi choose to opt out of national or global media networks or to utilize them for their own cultural, political, economic or educational objectives, their ability to participate in the life and evolution of The New Media Nation will be limited, unless access to the enabling technology improves. Levo-Henriksson observes that Native American radio stations are "important landmarks of Native American identity," and are "largely the result of a move away" from the assimilation that resulted from US government policy in the first half of the twentieth century (Levo-Henriksson 2007: 27).

Native American Public Telecommunications (NAPT) is a US based, increasingly global service based in Lincoln, Nebraska. Its flagship service is the American Indian Radio on Satellite network distribution system (AIROS), which operates 24 hours a day using the Internet and public radio. Supported by funding from the Corporation for Public Broadcasting, the Ford Foundation and a host of smaller foundations and donors, NAPT manages a website, e-mail access, and a video distribution service that includes an archive of Native American videos and public television programs available to tribal communities throughout the country. In sharp contrast to the broader US population's relative Internet wealth, Native Americans have had poor access to the Internet. As this situation slowly improves, Native American Internet services are expanding, and NAPT is no exception, having added MySpace and Facebook pages, blogs, and an online discussion forum.

KYUK, the first Native American-owned and operated radio station in the United States, opened in Bethel, Alaska in 1971, the only station broadcasting in the Yup'ik language. In 1973, it launched a television service, KYUK-TV and then a newspaper, the *Tundra Drum.* Indigenous stations are among those most injured by the Alaska government's diminishing commitment to public broadcasting. In the 1970s, Alaska Natives embarked on a communications revolution with the introduction of community radio into the isolated villages of rural Alaska. In part, these developments were the outcome of a general cultural renaissance among Alaska Natives. Indeed, the story of community radio in Alaska is most significant insofar as it is bound up in struggles for cultural control and revitalization among Alaska's diverse indigenous groups (Daley and James 2004: 139). Since 1977, the state has at least partially supported a system of satellite-delivered television to any community with a population over twenty-five. By 1990, the Rural Alaska Television Network (RATNET) was serving 248 communities (Daley and James 1994: 161).

In 2000, the Navajo Nation began a three-year process of installing a wireless network. Its first use was to implement the Local Governance Act (LGA), with funding from the Bill and Melinda Gates Foundation (Cullen 2005: 33). That private funding was needed is an indication of one of the main differences between Canadian First Nations communications and those in the United States, which receive almost no government support. Today, one of the network's main uses is to link teachers and students. There are also projects in which young people help to teach IT skills to elders (Cullen 2005: 34).

Recently, the Māori program of "language nests" has been taken up by the United States, and incorporated into the Esther Martinez Native American Language Preservation Act (H.R. 4766), which passed the US House of Representatives and Senate, and became law in 2007. It is unfortunate that neither the quoted text nor the editor of the normally exemplary Indigenous advocacy publication, *Cultural Survival Quarterly*, mentions the Māori precedent or gives proper credit to Māori for having originated the term and concept. The US bill is named for Tewa elder and language teacher, Esther Martinez, who died in September 2007. Though an important breakthrough, it is only a small beginning. It provides for a program that is far less comprehensive, and more watered down, than the Aotearoa program. Rather than having the language nests universally available, Indigenous communities in the United States will have to compete for grants to fund them (Cultural Survival Quarterly 2007: 5).

Those which do manage to get funded will be available to children aged seven and under, and their families, but the multigenerational plan imbedded in Māori society and language nests is less firmly entrenched. The US language nests will provide "a daycare-like environment with full language immersion." That emphasis suggests a more child-focused and a less extended family-focused program. The more age-segregated US approach is also seen in the plan to make the grants available for "elementary and secondary school students" (Cultural Survival Quarterly 2007: 5).

The Mashantucket Pequot language project

In 1638, speaking the Pequot language was an offense that could be "punished by beatings, being sold into slavery or death." The Treaty of Hartford, which ended the Pequot War, included clauses that all but eliminated the Pequot language and culture. Now, some 370 years later, the Mashantucket Pequot Tribe is promoting its language and hopes its younger members can use it to pass on their culture (Wallheimer 2006). "Through our children, our language will live," said Charlene Jones, Mashantucket Pequot Tribal Council secretary. When they started the language project in the mid 1990s, tribal members knew few words and there were no native speakers. To date, linguists and tribal researchers have found or reclaimed more than 1,000 words. "As a general rule, a language needs 50,000 words in order to be considered established." Mashpee Wampanoag linguist Jessie Little Doe has worked with the tribe to identify words from written documents and rebuild other lost words, using other Algonquian languages "as a blueprint to build Pequot words." She has had to work with documents written by Europeans in the seventeenth and eighteenth centuries. The tribe now offers language classes. "I was always taught our language was dead," said Joshua Carter, 29, of South Kingston, Rhode Island. Mohegan historian Melissa Tantaquidgeon Zobel said her tribe also is reconstructing its language (Wallheimer 2006).

ICC's Circumpolar Inuit Declaration on Arctic Sovereignty: New Ways of Forging Relationships among Governments and Indigenous Peoples

In April 2009, citing "a pressing need for enhanced international exchange and cooperation" in the light of states' "increasing focus on the Arctic and its resources," and the "easier access to the Arctic" afforded by climate change, the Inuit Circumpolar Council (ICC) adopted *A Circumpolar Inuit Declaration on Arctic Sovereignty*. It was announced from Tromsø, Norway, during the well-publicized Melting Ice conference (led by Nobel peace prize winner, Al Gore, and the Norwegian Minister of Foreign Affairs), and the Arctic Council foreign ministers' meeting. ICC Chair, Patricia Cochran, emphasized Inuit unity, the "unique relationships" between Inuit and each state government, and between Inuit and international structures and projects. "We are saying to those who want to use *Inuit Nunaat* [the Inuit homeland] for their own purposes, you must talk to us and respect our rights" (ICC 2009).

Duane Smith, Vice Chair for Canada said that it is "in the interests of states, industry, and others to include us partners in the new Arctic, and to respect our land claims and self-government agreements." Tatiana Achirgina, Vice Chair for Chukotka [Russia] sees the declaration as a foundation "to continue our self-government processes here in Chukokta in partnership with the Chukotka Administration and the Russian Federation." Vice Chair for Greenland, Aqqaluk Lynge, said it is "not an *Inuit Nunaat* declaration of independence, but rather a statement of who we are, what we stand for, and on what terms we are prepared to work together with others." Edward Itta, Mayor of the North Slope Borough and ICC Vice Chair for Alaska, noted Alaska's pivotal role in the founding of ICC, in 1977, and said: "This declaration will strengthen us as one people across four countries" (ICC 2009).

The declaration affirms that Inuit rights are supported by "international legal and political instruments and bodies such as the UN Charter; the International Covenant on Economic, Social and Cultural Rights; the International Covenant on Civil and Political Rights; the Vienna Declaration and Programme of Action; the Human Rights Council; the Arctic Council; and the Organization of American States. It declares that Inuit state-derived rights and responsibilities are not diminished under international law, and that sovereignty is itself "a contested concept" that evolves with changing governance models, such as the European Union. It notes that sovereignties can overlap, and are often divided within federations in ways that promote minority and Indigenous rights. It calls for inclusion of Inuit "as active partners" in "all national and international deliberations on Arctic sovereignty," and the building of "new partnerships with states." Its guiding principle is that development of natural resources "can proceed only insofar as it enhances the economic and social well-being of Inuit and safeguards our environmental security" (ICC 2009).

CHAPTER 3

Lessons from Canada

Amplifying Indigenous Voices

> The phone lines are down to the Yukon's most remote community. The people of Old Crow won't have any way of letting the Chief Returning Officer know who won in their riding, short of renting a plane and flying the results in. Those results could be crucial.
> —CBC Radio, Whitehorse (Alia 1991a)

The results were indeed crucial, and they arrived in a most unusual way. A "ham" radio operator picked up a message radioed from an airplane flying over Old Crow and relayed the information to Whitehorse (Alia 1991f). That convoluted, but effective, mode of transporting information may seem peculiar to those living in urban centers. In remote and northern regions, such occurrences are a part of daily life. Communication and transportation are inseparable; interdependence is not a theory, but a daily reality. Breakdowns in transportation and communication can mean life or death in places where radio or telephone lines link people with survival, as well as with each other. When it comes to sending and receiving news, Indigenous people are used to improvising.

Although "access" is often taken to mean simply the availability of technology, it is really about hierarchies of power. In Arctic Canada, as in other remote regions, people are informationally disadvantaged, and make extensive use of communication technologies to improve their access to information. In 1995, *YukonNet* joined the information universe. Yukoners who could now access the World Wide Web found some surprises, when the twice-weekly *Yukon News* went online.

> even though I live in a small community in a remote corner of the planet, somebody in New Zealand can read my local newspaper on the World Wide Web before I can get it from across the street. For information to flow like this in major urban centres may be commonplace; but here in Whitehorse, where people still talk to each other on the street ... the delivery of our community broadsheet via cyberspace seems absolutely Orwellian ... Perhaps the most salient benefit from the Internet's arrival here is not that we can access [the Web] but that the Web can access *us* (Killick 1996).

Indigenous Writing, from Early Days to the Present

Before the 1980s, "most books credited to Aboriginal authors were of the 'as told to' variety" (Twigg 2005: 7). After 1900, more than 170 Aboriginal authors and 300 books were published in British Columbia (BC) alone. Canada's "first Aboriginal-owned and -operated publishing company", Theytus Books, arrived in 1980. It was started by Randy Fred (Salish) and nurtured by the En'owkin Centre at Penticton, BC, with the involvement and support of author, Jeannette Armstrong (Twigg 2005: 9, 10). Many Indigenous writers in Canada are known equally on the US side of the border. Their multifaceted work crosses cultural and media boundaries, as well. A few examples:

> Martha Douglas Harris, a Métis of Cree, British, Creole, Scottish, and Guianese descent, wrote the *History and Folklore of the Cowichan Indians* (1901) and other works on Cree and Cowichan people. Emily Pauline Johnson, Mohawk, Anglican, and Quaker British from Six Nations in Ontario, widely known as a poet-entertainer and author of *Legends of Vancouver* (1912), underscored her bicultural heritage by performing half of each program in Mohawk dress and half in western-style evening dress (Twigg 2005: 25). George Clutesi was an artist, actor, broadcaster, and writer from the Tsde-Shaht First Nation at Port Alberni, BC, whose voice was heard on CBC radio broadcasts in the 1940s. In 1947, he wrote for the newspaper, *Native Voice.* He painted a mural for the Indian Pavilion at Montreal's Expo '67 and published stories and a study of the potlatch tradition. In the 1970s he also worked as a film actor (Twigg 2005: 61).
>
> Howard Adams (Métis) was a historian, university professor, and political activist (a leader in the "Red Power" movement) and President of the Saskatchewan Métis Association. Hum-ishu-ma (known as Mourning Dove) was Okanagan and Colville Salish from the western US and

Canada. Her collection of *Coyote Stories* was published in 1934 (Caldwell, ID: Caxton Printers, Ltd., 1934); her autobiography was not published until 1936, fifty years after her death (Twigg 2005: 41–43). Dan George was a writer, bass player, television and film actor. In 1970, he was nominated for an Academy Award for Best Supporting Actor for his portrayal of a Cheyenne chief in *Little Big Man*, for which he won the New York Film Critics Award and the National Society of Film Critics Award. His granddaughter, Lee Maracle (Salish/Cree), is a member of the Stø:lo First Nation and a well-known writer, academic, performer, editor, and storyteller.

Jeannette Armstrong is a distinguished historian, writer, and editor from Penticton, BC. Annharte (Marie Annharte Baker), Anishinabe from the Little Saskatchewan First Nation in Manitoba, is a poet, filmmaker, and co-founder of the Regina (Saskatchewan) Aboriginal Writers Group. Garry Gottfriedson of Kamloops, BC (Shuswap/Secwepemc) is a poet, horse breeder, and author of a history, *One Hundred Years of Contact* (1990), whose parents were activists in the Shuswap community in the days of George Manuel's leadership. Filmmaker Loretta Todd commissioned him to write the poem, *Forgotten Soldiers*, which she used as the centerpiece for a documentary on Indigenous War veterans (Twigg 2005: 122).

Barbara Hager (Scottish/Cree/Métis) was born in Edmonton, Alberta. A writer and broadcaster, she has lived in Seattle, New York, and Lexington, Kentucky, where she founded the Central Kentucky Writer's Voice program. She returned to Canada and wrote a thirteen-part documentary titled *Honour Song*; in 2002, CHUM Television hired her to co-host the arts and culture series, *The New Canoe*. She has her own production company, Arrow Productions. Her *New Canoe* co-host, Swil Kanim, comes from the Lummi (Salish) First Nation located just across the border in Washington State (Twigg 2005:143–144). Ernie Crey, executive director of the Stó:lo Nation's Fisheries program, is the former president of United Native Nations. With Suzanne Fournier, he co-authored *Stolen from our Embrace: the abduction of first nations children and the rebuilding of Aboriginal Communities*, which received the Hubert Evans Non-Fiction Prize in 1977. Tomson Highway (Cree, born near Brochet, Manitoba) is an internationally known playwright and novelist, and a former concert pianist. He won the Dora Mavor Moore Award twice, and was the first Aboriginal writer to receive the Order of Canada.

Figure 3.1. Spreading the word: the library at Moraviantown First Nation in southwestern Ontario, Canada. Photograph by Valerie Alia.

Journalism and Resistance

The linking of journalism and resistance is not new. Early newspaper editors and publishers were often jailed for their behavior in what are sometimes mildly called "disputes." Mobs smashed equipment, burned newspapers, and destroyed print shops. The violence coexisted with an at least superficially more benign pattern of symbiotic intimacy between journalism and government. At least five of the "fathers" of Canadian Confederation were journalists (Fetherling 1990: 40). In the 1860s, editors of newspapers in Toronto, Brantford (Ontario), and Montreal held cabinet posts. There are similar stories in other parts of the world. Although current standards generally support the principle of maintaining distance between news media and the political arena, that is not always the case, especially in northern regions and close-knit communities, where a handful of people hold public positions and move back and forth between journalism and government. In southern Canada and in national politics, there is at least a pretense of maintaining mutual independence. An exception was the 1994 federal election, when publisher Mel Hurtig founded the National Party and placed himself at the helm.

Newspapers arrived relatively late in Canada, following their emergence in Europe and the United States. They were usually founded by printers, and often began as newsletters, a pattern which continues today, with Indigenous community newsletters evolving into newspapers and other media. Early English-language "news books" emerged in London (England) in the 1620s (Steffens 1998). The first newspaper in North America, *Publick Occurrences Both Foreign and Domestick*, began publishing in Boston in 1690 (Fetherling 1990). The invention of the telegraph in 1844 transformed the distribution of news and information, Canada did not have its own wire service until 1917—due not to technological inadequacy, but to the Canadian railway's monopoly over dissemination of the news and ill-considered business decisions that created an absurdly circuitous system. Although direct transmission was now possible, Canada used the wires in a very different way: the Great North Western Telegraph system, an affiliate of the Grand Trunk Railway, sent information by wire, from station to station. As the news arrived in each railway station, it was translated from Morse code to English, transferred to paper, and carried by runner to each separate newspaper office. In 1858, came the first transatlantic cable and a communication revolution. That same year, North West (Rupert's Land) ended its status as an outpost of the Hudson's Bay Company and became a part of Canada, and the *Nor'-Wester* became the first newspaper in what was then the North-West Territories, and today is part of Manitoba. The paper was first published in Red River Settlement (now Winnipeg) and expanded to other communities. In 1869, its Fort Garry edition was sold to Dr. (later Sir) John Schultz, a non-practicing physician and merchant who distributed it by pony cart in summer and dogsled in winter.

Northern Broadcasting and the Aboriginal Communications Societies

Canada has long been the world leader in fostering broadcasting and film, in remote communities, and by and for First Peoples. Broadcasting has been more durable than print, and now is expanding via "new media". The leadership role persists, despite declining government funding, and increasing support in other countries. Northern and Indigenous broadcast services originated in other regions and were produced by outsiders.

Figure 3.2. Community broadcasting grants leaflet targeting Inuit in the eastern arctic, part of a government-funded initiative. The inside text is in English and Inuktitut syllabics.

Canada's national broadcaster, the Canadian Broadcasting Corporation (CBC), launched its Northern Service in 1958. More than twenty years later, a report commissioned by the national regulatory body, Canadian Radio-television and Telecommunications Commission (CRTC 1980), said that communications had a key role in preserving Aboriginal languages and cultures. It set the stage for a new era in Aboriginal broadcasting, not only in Canada, but throughout the world. Canadian programs and policies have continued to set international precedents and inspire Indigenous projects in many regions and countries. In 1983, Canada launched its Northern Broadcasting Policy and Northern Native Broadcast Access Program (NNBAP) and funded 13 northern Aboriginal Communications Societies (see below). Some were already in place and were incorporated into the new framework. Their mandate was to serve Aboriginal communities across Canada, with a focus on remote and under serviced areas. Under the umbrella of the National Aboriginal Communications Society (NACS) they became regional centers for production and distribution of radio, television, and print media. In addition, there are several Aboriginal Communications Societies that are not linked to NACS, in Alberta, Nunavik, and other parts of Québec, British Columbia, Nova Scotia, Saskatchewan, and Yukon.

Table 3.1. Aboriginal Communications Societies (Canada)

Name	Location	Founded	Media Outlets	Languages
Aboriginal Multi-Media Society (AMMSA)	Northern Alberta	1983	Windspeaker newspaper	English
			Radio and TV	Cree, English
Inuit Broadcasting Corporation (IBC)	Northwest Territories, Nunavut, Labrador, Nunavik (northern Québec)	1981	Radio and TV Television Northern Canada (TVNC) uplink launched in 1991; transferred to APTN in 1999	English Inuktitut
Inuvialuit Communications Society (ICS)	Beaufort district of the Northwest Territories	1976 Founded as Inuit Okangit Inumgun 1984 Reorganized in as Inuvialuit Communications Society	Tusaayaksat newspaper	English Seglit dialect of Inuvialuktun
			TV	Inuvialuit, English

Name	**Location**	**Founded**	**Media Outlets**	**Languages**
James Bay Cree Communications Society, Originally the Cree Regional Authority	Northern Québec	1973 started publishing	Cree Ajemoon newspaper	
		1990 TV	Radio and TV (now independent)	Cree
Missinipi Broadcasting Corporation (MBC)	Northern Saskatchewan	1984	Radio	Cree, Dene
Native Communications Society of the Western Northwest Territories (NCS) (Launched by the Indian Brotherhood of the NWT)	Yellowknife, NWT	1971 Network founded	Native Press newspaper, later the Press Independent	
			Radio	English, Dogrib, North Slavey, South Slavey, Chipewyan, Gwich'in (Loucheux)
Native Communications Incorporated (NCI)	Northern Manitoba	1971		
Mikisew Broadcasting Corporation, later incorporated into NCI		1984	Radio and TV	Cree, English
Northern Native Broadcasting, Terrace (NNBT)	Northern British Columbia	1985	Radio	English

Name	Location	Founded	Media Outlets	Languages
Northern Native Broadcasting Yukon (NNBY)	Yukon		Radio and TV	English, Southern Tutchone, Northern Tutchone, Swich'in, Kaska, Tlingit
Societé de communication Atikamekw-Montagnais	Northern Québec	1983	Radio	English Montagnais Atikamekw
Taqramiut Nipingat Incorporated (TNI)	Nunavik (Inuit northern Québec)	late 1970s	Radio and TV	Inuktitut
Wawatay Native Communications Society	Northern Ontario	1974	Wawatay News newspaper	English Ojibway-Cree
			Radio and TV	Cree Oji-Cree

Source: Updated from Alia 1999.

Continuity, Survival, Conflict Resolution: Radio and the Oka/ Kanehsatake "Crisis"

Despite increasing globalization, radio remains a medium of linguistic and cultural continuity, and sometimes of survival. Radio has been called the most grassroots of media. It is well adapted to oral cultures, and nomadic and remote community life. Where languages are threatened, Indigenous-language programming is often the main attraction, with talk radio providing a forum for social and political dialogue. Conway Jocks often spoke of the (inter)connectedness and intimacy this medium brings to people who live in small and/or remote communities. "Phone-in shows are lively extended family affairs." While most North American talk radio is AM, in Indigenous communities, "FM, easier and cheaper to build, is king, followed by trail radio, sometimes called 'moccasin telegraph.'" Talk-back radio "forges the communication links in ethnic neighborhoods, small towns and Aboriginal communities from the farthest Arctic coasts to the outskirts of major Canadian cities, sending hundreds of languages through the air" (Jocks 1996: 174). CKRK, the station he founded at Kahnawake, in 1978,

> began ... as the communication voice of the Kanienkehaka Raotitiohkwa Cultural Center, a local institution dedicated to the promotion and reinforcement of Mohawk culture and language. Establishing a forum for existing Mohawk-language speakers, it provided a support for the revitalization of the spoken Native word and the creation of a more vibrant cultural environment for ... students attending the band-controlled Mohawk immersion school system ... Its secondary objective was to inform non-Mohawk people, whom Jocks called "our drop-in listeners, of who we are and what we believe in." (Roth 1993 [online version, unpaginated])

Along with linguistic and cultural survival, radio can also foster physical survival, as in the case of the 1990 "Oka Crisis." Townspeople at Oka, Québec sought to extend a golf course onto a sacred Mohawk burial ground. Mohawk resistance and police and military intervention led to a blockade of Kanehsatake. To assure continuing coverage, Mohawk broadcasters Marie David and Bev Nelson camped out in radio station CKHQ, where they remained until the crisis ended four months later. Throughout the history of Indigenous broadcasting, some of the most effective communications strategies have been low-tech. When Marie David and Bev Nelson found themselves trapped inside their radio station, living on rationed food supplies and sleeping in the station, the Internet was not yet an option. Yet they managed to create a network that reached across their own community to other First Peoples, and ultimately, internationally. The telephone "made it possible, in a sense, to be in two places at the same time" through "party lines" on which conversations could take place in many homes, simultaneously. In "the early systems bells rang along the entire line and everyone who was interested could listen in" (Roth 1993 [online version, unpaginated]). Using the radio station's single telephone and fax line, Nelson and David kept their relatively isolated community connected to individual and group supporters and others, and brought their eye-witness account to the world. As other radio stations within and outside Canada picked up their broadcasts, the listening audience rapidly grew. They were interviewed almost daily by Canadian and international radio and television services. Excerpts from CKHQ reached other stations by telephone and occasionally, were "sneaked out by a supportive Montreal journalist for rebroadcast" on stations in Montreal, across Canada, and internationally (Roth 1993 [online version, unpaginated]).

Nearby, the larger community of Kahnawake provided important support. Unlike Kanehsatake, it was not occupied by Canadian authorities, and thus had more freedom to operate. Led by its founder and station manager, Conway Jocks, Radio Kahnawake CKRK "became the loudest First Nations broadcast voice in southern Québec," reaching more than 300,000 listeners during peak listening periods and playing "a pivotal role in providing alternative forms of information," by building "a public opinion support base for the Mohawk position and acting as a conflict mediator." At the center of the service was CKRK's phone-in show, *The Party Line* (Roth 1993 [unpaginated]).

During the confrontation, CKRK's policy was to act as "normal" as possible, though music playlists "tended to have high message value, e.g., *Give Peace a Chance* by John Lennon or *The Freedom Song* by Frosty, a well-known resident of Kahnawake." The programming featured surveillance information; tips on surviving with limited resources; public service announcements of military and police maneuvers; "the comings and goings of residents"; updates on political negotiations; appeals from members of the Band council to maintain calm and sobriety; suggestions of how to answer public questions; conversations with witnesses and participants from the front lines; and advertisements and political statements. The station became one of the most important vehicles for keeping the town informed and keeping the channels to outside communities open. Roth maintains that talk-back radio "was the most effective way of encouraging and promoting communication with outsiders"—a "potent medium for producing public debate" that opens up a safe mediaspace in which audiences can join the discussion while remaining anonymous. The discussion was limited only by the judgment of CKRK's directors and broadcasters, and the constraints imposed by Canada's *Broadcasting Act* (1991) and *Radio Regulations* (1986), which proscribe transmission of racist comment (Roth 1993 [unpaginated]).

Believing that, during a conflict, rumor is the most dangerous incendiary device, Conway Jocks insisted on confirming the facts before reporting them, and declined to broadcast what could not be confirmed—a departure from the style of talk-back radio experienced elsewhere. CKRK could not afford to join the BBM (Bureau of Broadcast Measurement) to obtain official ratings, but their "mole at one of Montreal's major radio stations told us that we were pulling in almost half a million listeners during our phone-in segment. If true, that is the biggest story of all—and the best kept secret." The first phone-in broadcasts had such large audiences and were so important to helping to defuse conflict that Jocks decided to extend *Party Line*'s broadcast hours and bring in a former announcer, Nathalie Foote. "The program soon turned into a barricade jumper, the only direct link with the outside world for us inside who were rapidly taking on a fortress mentality" (Jocks, quoted in Roth 1993).

In the non-Indigenous world, talk-back radio is known for its provocative and confrontational style. In Mohawk country, it took a diametrically opposite position. During the summer of 1990, each *Party Line* broadcast opened with a prayer:

> Great Spirit, whose voice I hear in the winds and whose breath gives life to all the world, hear me. I come before you—one of your many children. I am small and weak. I need your strength and wisdom. Let me walk in beauty and make my eyes ever behold the red and purple sunset. Make my hands respect the things you have made, my ears sharp to hear your voice. Make me wise so that I may know the things you have taught my people, the lesson you have hidden in every leaf and drop. I seek strength not to be superior to my brothers, but to be able to fight my greatest enemy—myself. Make me ever ready to come to you with clean hands and straight eyes so when life fades as a fading sunset; my spirit may come to you—without shame.

The prayer was followed by the rules of conduct, which prohibited foul language, use of last names, and naming of people entering or leaving Kahnawake. Roth thinks Nathalie handled crisis radio well, making good use of her friendly and "chatty" manner. Coping with more than 100 calls an hour some of the time, she spoke in a soft voice, inviting callers to share their feelings, even encouraging expressions of racism, deflected from violent action to (angry and offensive) speech—deflected from the potential for direct and personal harm—and Mohawk "effigy-burners and rock-throwers" to call in and explain their actions—again redirecting violence to talk. She was influenced by Martin Luther King's Gandhian view that "if violence is the language of the inarticulate," the opportunity to speak can provide an outlet for anger and frustration, and can diminish physical violence. When callers made hostile comments, Nathalie quietly thanked them and hung up. She rarely "blew her cool" and managed to maintain diplomacy "through some very tense periods. Nat developed new ways of using radio to diffuse hostile energy." Using radio "for catharsis and conflict mediation ... she developed a technique for diffusing tension—psychotherapeutic radio" (Roth 1993).

The role of CKRK-FM at Kahnawake, and the inside-the-barricades CKHQ broadcasts by Marie David and Bev Nelson at Kanehsatake, echoed the earlier experience of Lakota people during the 1973 siege at Wounded Knee in South Dakota. During that crisis, Lakota women occupied, and where necessary, camped out at their community radio station and kept the broadcasts going under tense and sometimes life-threatening circumstances (Crow Dog 1991).

Figure 3.3. Banner in "The Pines," at the first annual spiritual gathering and traditional pow wow, Kanehsatake, Mohawk Territory (Québec), 1991. Photograph by Valerie Alia.

Figure 3.4. The much-loved Kanehsatake elder, Walter David, Sr., carrying the Mohawk flag at the 1991 gathering. Among his children are the journalists Dan, Valerie, and Marie David, whose radio activities during the "Crisis" are described above. Photograph by Valerie Alia.

Broadband, Mobiles, and Digital Technology

A 2007 study in Britain showed digital platforms to comprise about 16 percent of all radio listening and digital radio or DAB—digital audio broadcasting—to be the preferred vehicle for receiving radio broadcasts, with digital television and the Internet following close behind (Plunkett 2007: 3). Keith Battarbee notes the emerging potential of mobile (cell) phones and digital radio to enhance the survival of Indigenous languages (in Chapter 4, we look at other uses of mobile telephony, in Africa). Battarbee's study of the "impact of information and communications technologies (ICT) on language use identifies two factors that influence heritage language use and maintenance that 'significantly changed' in the late twentieth century"—the "drastically enhanced respect accorded aboriginal cultures" and "the advent of electronic technologies for communication and for the sharing of information." He finds that telephony "is essentially language-neutral, and the accessibility, coverage and annihilation of distance provided by cellular and satellite networks and by voice-over-internet protocol thus potentially represents a significant resource supporting minority languages." He points to hazards as well as benefits—the continuing bias, in computer technology, towards written language and domination by English, along with the "rapid growth of other-language websites and other resources" (Battarbee 2007).

In 2006, Industry Canada announced its Broadband for Rural and Northern Development Pilot Program, which offered up to $542,723 in program savings "available to be invested in the Kittiwake Economic Development Corporation" to provide broadband, or high-capacity Internet, to communities not currently receiving broadband service (Industry Canada 2006). This next generation of technological support follows Canada's long-time commitment to serving rural, remote, and Indigenous communities. Within a short time after its inception, the Program had extended its reach to "almost 900 communities, including 142 First Nations reserves." Combined with the $155-million National Satellite Initiative, "Broadband service to rural and remote communities will ultimately provide Aboriginal, northern and rural communities with more opportunities to move forward socially and economically" (Industry Canada 2006: 1).

Triumphs and Disappointments in Southwestern Ontario

In 1990, a group of Indigenous and non-Indigenous journalists, writers, educators, and volunteers organized the Native Writers' Circle at N'Amerind Friendship Centre in London, Ontario. The focus shifted from appreciating talent, to providing paid work for Indigenous journalists, and the group renamed itself Native News Network of Canada (NNNC). The writer and broadcaster, Bud (Enos) Whiteye, (Delaware) became its first President. It had two "homes", in N'Amerind, and at Bud's home community of Moraviantown First Nation. It was intended that NNNC would become a national news service—an alternative to mainstream wire services that would market stories by Indigenous journalists to a variety of media outlets. Although a few stories were produced and sold, to newspapers and broadcast outlets, NNNC never became a full-fledged network. Its second President, Dan Smoke, took it in a different direction, developing what was originally a secondary project in community radio into a weekly program, *Smoke Signals First Nations Radio.*

With donated studio space, airtime, and an office at the University of Western Ontario (UWO), the project was relocated on campus, and in Dan Smoke's home community of Ohsweken (Six Nations). NNNC faded into the background; *Smoke Signals* has thrived. Co-hosted by Mary

Lou Smoke (*Asayenes Kwe*) (Ojibway) and Dan Smoke (*Asayenes*) (Seneca), it provides a forum for discussion of a wide range of issues and topics and an outlet for Indigenous music. A few years ago, Mary Lou Smoke formed a women's network (Buddle 2008). There are also television broadcasts—"First Nations on the Evening News Report," on London Ontario's mainstream channel, CFPL, and on cable television's "A Channel"—"southern Ontario satellite TV. We're reaching eight million people," Dan Smoke told me in a telephone conversation, adding that "we also do 'Interstitials'—mini-features on a commercial slot," maintain websites and a blog, and teach a course on First Nations and media at the university (Smoke 2006). The course continues a process begun in 1990, when Bud Whiteye and I developed the First Nations Intensive seminar for students in UWO's Graduate School of Journalism, a day-long "teach-in" aimed at encouraging future journalists to report more accurately and sensitively on Indigenous issues. The seminar ran for several years, received enthusiastically by Indigenous communities and students, but with grudging and diminishing support from the Journalism school, at a time when the Programme in Journalism for Native People was facing closure. The Smokes were among the guest lecturers who brought expertise from several fields, communities, and cultures. Despite its continuing media/journalism focus, their new course seems to have found a more congenial home in the Anthropology department.

Over the years, there were expressions of disappointment that the original objectives, of providing paid work for professional Indigenous journalists and disseminating information nationally, were never met. Some Indigenous and non-Indigenous volunteers left in frustration. One long-time volunteer said,

> NNNC has never become real as a network. They do important work, but have not really seemed motivated to create a network. There really is no NNNC; it is mainly *Smoke Signals* (Alia personal communication 2006).

Smoke Signals energetically serves First Nations and other people on the main UWO campus, in much of the city of London, and in nearby First Nations communities, with the website and television broadcasts expanding its range. It has been highly successful as a community-building, public information and community action project.

In 1997, I was asked to write a statement of principles for NNNC, in consultation with its president and directors, which would reflect its members' priorities and ethical concerns. We reviewed hundreds of codes and statements from journalism organizations around the world. In some respects, it reflects other codes and statements of principles. In other ways, it represents a significant departure from conventional journal-

Figure 3.5. Bud (Enos) Whiteye, founder of Native News Network of Canada and co-founder of the First Nations Intensive seminar, at his home community of Moraviantown (Delaware) First Nation, Ontario. Photograph by Valerie Alia.

istic practice—for example, in seeing the journalist as a citizen and member of the community rather than as a politically pure, neutral bystander. The complete text is included, in the Appendix, *Native News Network of Canada Statement of Principles.*

Political and Media Leaders

In all of the countries I have known and observed, there has been a tendency for Indigenous leaders to spend part of their careers in journalism. This is especially true in Canada, where a number of the most prominent leaders started in broadcasting. Among them are Jose Kusugak, who was heard on BBC and IBC before moving to Inuit politics; and Mary Simon, Rosemarie Kuptana, and John Amagoalik, whose careers are sketched below.

The appointment in 1994 of Mary Simon as Canada's first Ambassador for Circumpolar Affairs marked the first time an Inuk would hold an ambassadorial position in Canada. Like many Indigenous leaders, she began in journalism, broadcasting on CBC Northern Service and writing for *Inuit Today*. She switched to politics in 1973, and from 1986 to 1992, served as the first President of the Inuit Circumpolar Council (then called the Inuit Circumpolar Conference), where she developed close ties to Greenland, sought "a significantly expanded role for Inuit at the international level" (Petrone 1988:272) and supported ICC's push for international cooperation and designation of the Canadian Arctic as a nuclear-free zone—a position Inuit continue to promote despite challenges from governments and corporations. Her opposition to US President Reagan's "Star Wars" Strategic Defense Initiative is timely, in light of Russian and US initiatives to expand their presence in the Arctic (Petrone 1988: 265–271). Simon attributes her early awareness of the internationality of Inuit to hearing Greenlandic music on her family's short-wave radio as they traveled among hunting and fishing camps in northern Québec, and to her grandmother's dream that Inuit from different countries would someday work together. As Ambassador for Circumpolar Affairs, her main responsibility was to help develop the Arctic Council. At its inaugural First Ministerial Meeting held in Iqaluit in 1998, the council issued the "Iqaluit Declaration: An Agenda for 2000." It called for circumpolar cooperation, an initiative on children and youth, and a northern foreign policy for Canada.

Rosemarie Kuptana is an internationally prominent Inuvialuit broadcaster, political leader, environmental and political activist, negotiator, and consultant. She was the first woman to serve as President of the Inuit Broadcasting Corporation (IBC). Her traumatic childhood experience in a government residential school inspired a lifelong determination to resist assimilationist programs and policies. From the mid 1970s onward, she participated in land claims discussions between Inuit and the Canadian Government. In 1979, she joined CBC Northern Service as host of CBC Inuvik's radio public affairs magazines. She moved on to IBC and as President, strengthened its commitment to developing Inuktitut- and Inuvialuktun-language programming that promoted Inuit traditional life. She was a key member of the group of Indigenous communications leaders who envisioned, designed, and developed Television Northern Canada (TVNC).

In the early 1980s, Kuptana left broadcasting for politics, becoming Canadian Vice-President of the Inuit Circumpolar Conference (ICC) from 1986 to 1989, and from 1991 to 1996, President of the Inuit Tapirisat of Canada (ITC). She was the principal negotiator for Inuit in the process by which Aboriginal people attempted to win constitutional recognition of their inherent right to self-government. Despite strong lobbying and valiant efforts at the negotiating table, the effort failed—a permanent sore point for those who consider the Nunavut Agreement a diluted fragment of the agreement they had originally sought. She had greater success in negotiating an amendment to an international treaty that recognized and constitutionally protected Inuit hunting rights.

Her predecessor as ITC President was the broadcaster, political leader, and visionary, John Amagoalik, who is often called the "father of Nunavut" and was a key player in all of the developments leading up to the creation of Nunavut, and the development of TVNC and APTN. John Amagoalik has always excelled at seeing the big picture. Declaring that there is more to literacy than reading, and more to "reading" than the printed page or computer screen, he eloquently clarifies the cultural core of literacy, while unraveling the layers of colonial thinking:

> If we are to survive as a race, we must have the understanding and patience of the dominant cultures of this country. We do not need the pity, the welfare, the paternalism and the colonialism which has been heaped upon us over the years. We must teach our children their mother tongue. We must teach them what they are and where they came from. We must teach them the values which have guided our society over the thousands of years ... our philosophies which go back beyond the memory of man. We must keep the embers burning from the fires which used to burn in our villages so that we may gather around them again. It is this spirit we must keep alive so that it may guide us again in a new life in a changed world (Petrone 1988: 210).

Television Northern Canada (TVNC)

> At exactly 8:30 p.m., an Inuktitut voice signals the start of the world's largest aboriginal television network. Elder Akeeshoo Joamie of Iqaluit asks Jesus to guide TVNC to success. An English translation rolls slowly across the screen ... The vision of TVNC became a reality with a montage of Inuit, Dene, Metis, Gwich'in, Kaska, Tuchone, Tlingit and non-aboriginal faces beamed to 22,000 households from Northern Labrador to the Yukon-Alaska border ... TVNC is a non-profit consortium which aims to use television for social change. —Thomas 1992: 14

As its founders saw it, "Northern aboriginal television programming developed as a front-line response to the invasion of foreign television signals entering the North in the late 1970s" (TVNC 1998: 1). In the late 1970s and early 1980s, Canada's Department of Communications responded to extensive lobbying by Inuit organizations by developing the Anik B trial-access program, which comprised a series of interactive audio and video experiments. In 1976, the satellite carried an experimental interactive audio project across northern Québec (now Nunavik). The project, called Naalakvik I, linked eight radio stations and was run by the Aboriginal Communications Society Taqramiut Nipingat Incorporated (TNI) (Roth and Valaskakis 1989: 225). In 1978, the Anik B satellite carried the launch of the programs initiated by Project Inukshuk, the media project named for the human-form stone sculptures Inuit use to mark important features and places on the land. The federally funded project heralded the start of Inuit-produced television. The Inukshuk project was sponsored and organized by Inuit Tapirisat of Canada (ITC, now ITK), with video production facilities in Frobisher Bay (now Iqaluit) and Baker Lake. "The purpose of Inukshuk was to train Inuit film and video producers ... establish Inuit production centres in the North, and ... conduct interactive audio/video experiments utilizing the 12/14 GHz capability of the satellite to link six Arctic settlements" (Roth and Valaskakis 1989: 225). When Inukshuk first went to air in 1980, it sent 16-1/2 hours a week of television programming and teleconferencing to the six communities.

From the start, it was known that these were short-term projects that would require longer-term funding to survive and develop. In 1980, the CRTC formed a nine-member committee to consider proposals to develop satellite television services for northern and remote communities. It was headed by Réal Thérrien and included John Amagoalik, who would become the first Indigenous person to help set national communications policy in Canada. The committee emphasized the role of broadcasting in preserving and maintaining Aboriginal languages and cultures, and identified the emergence of "A New Broadcasting Universe":

> Our first unanimous conclusion is that immediate action must be taken to meet the needs of the many Canadians who believe that, as regards broadcasting, they are being treated as second-class citizens ... We cannot stress too strongly the immediacy of the problem: alternative television programming must be provided from Canadian satellites with no further delay (Thérrien 1980: 1).

The authors of the Thérrien Report wrote that the new satellite-delivered television posed both "the most damaging threat to native objectives and the most potentially feasible means of achieving them" (TVNC

1998: 2). Ken Kane, of Northern Native Broadcasting Yukon, was TVNC's first Chairperson. He said,

> When I first heard about this new technology coming to the North, I realized that along with it would come a lot of change and a lot of impact for my people. That is why we got involved. To make sure that this time we got in on the ground floor: not to oppose it or go against it, but to grow with it. To learn and develop with it (TVNC 1998: 2).

In 1991, the world's largest Aboriginal television network was born. TVNC's mandate was to broadcast "cultural, social, political and educational programming" in English, French, and several Indigenous languages, to Canada's northern Indigenous people, via satellite. Its leaders represented the 13 Aboriginal communications societies (See Table 3.1) and a handful of associate member organizations. TVNC served 100,000 viewers north of 55 degrees latitude, covering one-third of Canada (TVNC 1993). In its new incarnation as APTN, it broadcasts nationwide. TVNC was a precedent-setting experiment that succeeded beyond its founders' expectations, using Canada's pioneering Anik communications satellites to transmit its programming to an increasingly widening audience.

In 1998, TVNC produced a position paper. The following statement is from a key section titled "Facing the new era":

> Aboriginal peoples in the North are again at a crossroads. Digital video compression (DVC), universal addressability, direct-to-home satellite services (DTH) and the expansion of the information highway have [changed] and will dramatically change business, information and cultural environment. The paradox identified in the Therrien Report is still true today ... Aboriginal participation in the design, ownership and operation of communications services ... will strengthen the foundation of the northern communications infrastructure and ensure that the new services support the practice of self-government (TVNC 1998: 4).

TVNC's founders saw this and other media and communications projects as inherently and importantly linked to the emergence of Indigenous self-government, (on- and off-line) education, and cultural and political survival. Their position paper reminded the government of this. "The inherent right to self-government places aboriginal people in control of all aspects of community life. Programs which were historically administered by the Department of Indian Affairs are being devolved to aboriginal governments." Most importantly, "Access to information is essential to both effective self-governance and the economic develop-

Figure 3.6. Cover of the press kit distributed at the launch of Television Northern Canada (TVNC).

Figure 3.7. TVNC: "The Dawn of a new Era." Part of the initial publicity campaign promoting Television Northern Canada.

ment necessary for aboriginal communities to become self-sufficient" (TVNC 1998: 6). On 1 September 1999, after several years of successful broadcasts, TVNC received permission from the CRTC to become a national service—the Aboriginal Peoples Television Network (APTN).

One of the most popular programs on TVNC was *Super Shamou*, an amalgam of Inuit and Qallunaat characters and traditions, with distinctively Inuit qualities and values. The following images are from the English and Inuktitut language versions of the Super Shamou comic book that was designed to accompany the show. "The Origin of Super Shamou" echoes the familiar Superman legend. The portrayal offers both a send-up of Superman and a uniquely Inuit character. *Super Shamou* gave birth to another popular TVNC children's program, *The Takuginai Family*, which taught "respect for elders, traditional Inuit culture, and the Inuktitut language" (David 1998: 37). *Takuginai* means *Look Here* in Inuktitut. These characters and programs carried over, and evolved, on APTN.

The Aboriginal Peoples Television Network (APTN)

APTN is run and produced by Aboriginal people for a nationwide, multicultural audience who receive it as part of their basic cable television package. APTN's 2008 website and brochure proclaimed:

> Aboriginal Peoples Television Network (APTN) is the first national Aboriginal television network in the world with programming by and about Aboriginal Peoples. APTN brings a full slate of culturally diverse, entertaining and intelligent programming to all Peoples [with] an emphasis on the development of Aboriginal talent. APTN is available to approximately 10 million Canadian households and commercial establishments with cable, direct-to-home satellite (DTH), telco-delivered and fixed wireless television service providers (APTN 2008a).

In 2008, 28 percent of APTN's programs were in Aboriginal languages (up from 25 percent in 2006); 16 percent were in French (up from 15 percent in 2006); and 56 percent were in English (down from 60 percent in 2006) (APTN January 31, 2006; APTN 2008).The range of languages is extended by subtitles that are included with some of the programs. The network shuns government funding, surviving on "subscriber fees, advertising sales and strategic partnerships" (APTN 2008). The broadcast day comprises mainly independently produced programming, along with three national news shows, produced in-house and broadcast from APTN headquarters in Winnipeg. The newscasts

Figure 3.8. Super Shamou comic book, designed to accompany the program aimed at children and teenagers, featuring an Inuit superhero, one of the most popular offerings on IBC and TVNC. Front cover: Inuktitut version. The comic was also published in English and French.

Figure 3.9. Transitional advertisement from TVNC, announcing the inauguration of APTN.

are supported by "in-studio interviews and reports from 11 news bureaus across the country telling stories from an Aboriginal perspective" (APTN 2008). Following the long tradition of community participation and engagement in Indigenous radio, there is also a weekly, live call-in show, "Contact." On radio and television, Indigenous call-in shows bear little resemblance to those seen and heard on mainstream networks.

The difference in production values was brought home to me in the early 1990s, in Whitehorse, Yukon. I was in the studios of NNBY, observing a live broadcast of the radio phone-in show, during the "Oka Crisis." Callers had important, and often emotional, things to say. The show was scheduled for a one-hour time slot, but as the hour ended, the phones were still ringing, with no sign of letup. In the space of a few minutes, a decision was made to continue the show, immediately following the station's remote/spot coverage of a healing circle and demonstration expressing solidarity with Mohawks at Kanehsatake. Following the demonstration, the camera crew returned to the studio and resumed broadcasting the phone-in show, which continued until all of the callers had an opportunity to speak.

Community programming has that kind of flexibility. While providing alternative perspectives and ways of interviewing, editing, and presenting, as a national broadcaster, APTN maintains a less flexible, more fixed schedule. In keeping with the principle of tradition-plus-innovation, APTN has a free membership portal and news streaming on its website.

To celebrate the APTN launch, the October/November 1999 issue of *Aboriginal Voices* magazine ran a cover photo of APTN's first chair, Abraham Tagalik (a veteran of IBC and TVNC) headlined: "Please Adjust the Colour on Your Set." In Whitehorse, on 10 September 2000, just after APTN's first anniversary, I tuned in midway through an Inuit current affairs magazine featuring mini-documentaries on northern Nunavut Day celebrations and a skidoo race captured by energetic and creative camera work.

In 2006, the newly elected chair of APTN's Board of Directors, Judy Gingell, said the Directors "share the vision of taking APTN to the next level, which is to become the digital gathering place of Aboriginal Peoples around the world" (APTN, 26 January 2006). Along with the mandate to provide an Indigenous "digital gathering place" is an objective of educating and informing non-Indigenous people—seen not only in APTN, but also across much of The New Media Nation. An audience survey showed that, although APTN's primary audience consists of Aboriginal Peoples, "current audience measurement of sampled homes in large urban settings primarily captures the secondary [non-Aboriginal] audience … and these viewing numbers have grown … over the last two years APTN has seen its average weekly reach grow from 1,563,000 viewers in the period of January 2004 to October 2004 to 2,115,000 viewers in the same period one year later, a growth of 35%" (APTN 31 January 2006).

APTN's broadcast week includes outstanding music programs, such as "Rez Blues" and "The Mix". In 2008, APTN expanded its commitment to Indigenous musicians by cosponsoring a new national distribution program aimed at bringing Indigenous music to the mainstream market. APTN's partners are Errol Ranville and his cweedband.com, NewCap Radio, a syndicate of more than 80 stations, and a company called CDPlus. The APTN website quoted John Toews of CDPlus, who said, "This is an exciting chance to provide visibility and a dedicated platform to a group of musicians typically under-represented in today's marketplace" (APTN 27 August 2008).

Figure 3.10. Early promotional piece for The Aboriginal Peoples Television Network (APTN).

In 2008, the Conservative government reluctantly and belatedly followed Australia's lead and made an official apology to Indigenous people in Canada.

> For the sexual and physical abuse that occurred at the schools, Canada apologized. For the efforts to wipe out aboriginal languages and culture ... Mr. Harper expressed remorse. Unlike Australian Prime Minister Kevin Rudd in his apology to aboriginals in February, Mr. Harper made no promises to improve aboriginal social conditions.
>
> When the politicians had finished, five aboriginal leaders ... seated in a circle ... with six former residential school students were given the unprecedented opportunity to ... speak ... The Prime Minister's message of healing was undercut ... by one of his own MPs, [Pierre Poilièvre] who spoke sarcastically of the size of the residential schools settlement on talk radio (Curry and Galloway 2008 unpaginated).

US media are notoriously inattentive to Canadian news. This occasion was an exception. The *Washington Post* reported:

> Canadian Prime Minister Stephen Harper delivered a long-anticipated apology yesterday to tens of thousands of indigenous people who as children were ripped from their families and sent to boarding schools ... part of official government policy to 'kill the Indian in the child.'... thousands of Indian, Inuit and Metis children suffered mental, physical and sexual abuse in 132 boarding schools, most of them run by churches [from the late 1800s to 1996]. "The treatment of children in Indian residential schools is a sad chapter in our history," said Harper, facing indigenous leaders who sat in a circle in the House chamber [and] listened silently or wept for what their people suffered and are still suffering (Brown 2008: A01).

ITK President Mary Simon told the House of Commons: "Let us not be lulled into believing that when the sun rises tomorrow, the pain and scars will be gone. They won't. But a new day has dawned." Some years ago, when Simon was the international ICC President, I was commissioned to write an article about her for a major Canadian magazine. When the piece was cut to an insulting single paragraph, I consulted Simon, and withdrew it. As is so often the case with mainstream media, the magazine missed an opportunity for news leadership. In 2008, the media finally paid attention. The *Washington Post* report included information about the historic 2006 settlement, in which the government agreed to pay "80,000 former residential school students $10,000 each for the first year they attended the schools and $3,000 for each subsequent year [with] additional compensation for sexual and physical abuse ... the largest class-action settlement in Canadian history." Also included in the settlement was a commitment to establish a truth and reconciliation commission, "the first of its kind in an industrialized country" (Brown 2008: A01). Other such commissions have been organized in South Africa, Argentina, Peru, and Sierra Leone.

Harry S. LaForme, a justice of the Ontario Court of Appeal and member of the Mississauga First Nation, heads the commission, which began its work in June 2008. In an interview, he said that the "policy of the Canadian residential schools wasn't to educate Indian children ... it was to erase the culture of Indian people from the fabric of Canada" (Austen 2008).

"The memory of residential schools cuts like merciless knives at our souls," said Phil Fontaine, national chief of the Assembly of First Nations [AFN], which represents 633 Indigenous communities across Canada. "[Fontaine] is widely credited with being the first aboriginal person to go public about abuse he experienced in a boarding school ... 'I was one of the people who suffered physical abuse as well as sexual abuse. Sadly, I am not unique.'" Kathleen Mahoney [AFN chief negotiator] said in a telephone interview that many children succumbed to tuberculosis. "In some schools, 50 to 60 percent of the children died. ... Deaths went unreported. Many were buried in unmarked graves. Many children ran away and died because of drowning or freezing" (Brown 2008: A01). Duncan Campbell Scott, deputy superintendent of Indian Affairs from 1913 to 1932, set the tone for decades of abuse in his contribution to a government document, expressing his policy objectives: "I want to get rid of the Indian problem ... Our objective is to continue until there is not a single Indian in Canada that has not been absorbed into the body politic" (quoted in Brown 2008 and others).

The US media were not just reporting on Canada. They were pointing inky fingers at their own national policy. In February 2008,

> The US Senate passed a resolution that apologized for atrocities committed against Native Americans as part of the assimilationist policies, appropriation of Indigenous lands, and forcible removal of children from their families to distant boarding schools. The resolution urged President George W. Bush to acknowledge and apologize for the mistakes; as of August 2008, the resolution was still awaiting action in the House of Representatives, and no apology had been issued (Brown 2008: A01).

Some media got it partly wrong—"Canada tells its Indians: We're sorry," headlined a story published in the *Chicago Tribune* and, under a more accurate headline referring to "Canada's Native People," in the *Los Angeles Times* (Guly and Farley 2008). By contrast, a Canwest News Service story published in several Canadian papers referred to "First Nations, Inuit and Métis peoples," but its unequivocally celebratory tone suggests that the journalists may have listened selectively, and rather narrowly, to Indigenous people. "Making parliamentary history by speaking on the floor of the House of Commons, representatives ... welcomed the apology" (O'Neill and Dalrymple 2008: A12). Others were less celebratory:

> The apology was billed by the government as a chance to redress a dark chapter in Canadian history and to move forward in reconciliation. But the hours before the landmark statement were marked by wrangling over whether native leaders were adequately consulted ... and anger that they would not be allowed to respond in the House of Commons. Just before Harper's speech, opposition leaders led a successful motion to allow aboriginal representatives to reply in the chamber. Some survivors ... said the apology came only grudgingly under intense pressure from native groups, and must be matched by action (Guly and Farley 2008).

Where Australian Prime Minister Kevin Rudd had *campaigned* on his promise to apologize to Indigenous people, and issued an apology soon after his election, Canadian Prime Minister Stephen Harper was *pressured* and shamed into his apology, first refusing and then stalling before finally delivering it. More than a year earlier, Liberal MP Tina Keeper posted the following comments on her website:

> The Minister of Indian Affairs must honour the federal government's pledge to Canada's First Nation, Métis and Inuit peoples and issue a formal apology to residential schools survivors, Liberal Opposition MPs said today. Yesterday, Indian Affairs Minister Jim Prentice and the Conservative government confirmed that the settlement for survivors of Indian residential schools would not include an apology. "The Minister's refusal to issue a simple human apology to residential school survivors is another broken promise and betrayal of Aboriginal Canadians," said Desnethé-Missinippi-Churchill River MP Gary Merasty (Keeper 2008).

The Minister's comments contradict a written pledge made by the Assembly of First Nations and former Liberal government and signed by then-Deputy Prime Minister Anne McLellan. "It is amazing that the Harper government doesn't have the decency or compassion to live up to the promise to issue an apology," said Labrador MP Todd Russell. "An apology would cost nothing, but it would go a long way towards ending this terrible chapter in Canadian history. It is shameful that the Government will not take this simple but important step to close the residential schools legacy," said Liberal Indian Affairs Critic Anita Neville (Keeper March 2007). "Mr. Harper and many fellow members of the Conservative Party initially resisted offering an apology." "[Assembly of First Nations Chief Phil] Fontaine said in an interview that he believed that

Mr. Harper changed his mind after the government of Australia formally apologized to its aboriginal people earlier this year for its policy of forced assimilation" (Austen 2008 [online, unpaginated]).

Before being elected MP for Churchill Manitoba, Tina Keeper was better known as an actor of considerable talent. She starred as an Aboriginal RCMP officer in the television series, "North of Sixty." One particularly moving episode recounted her personal struggle with the effects of her residential school experience. As a politician, she played a major role, for several years, in efforts to pressure the Harper government to apologize to Canada's First Peoples. Yet her considerable influence and leadership has been written out of much of the political history making and the media coverage. Even in its final agreement to apologize, the Harper government continued to deliver insults to Indigenous people. It is, I fear, a clue to Harper's truer attitudes—that he acceded to the same request when it was made by New Democratic Party (NDP) MP Jack Layton. Ms. Keeper is Aboriginal. Mr. Layton is White. The insult was underscored by Harper's public expression of thanks to Jack Layton and failure even to acknowledge Tina Keeper. The day before the apology, Harper rejected a request from Keeper to include Indigenous people in the proceedings. He then acceded to virtually the same request, when it came from Layton—and during the apology, singled out Layton "from among the three opposition parties for helping make the apology happen. The native leaders were told only an hour before the apology that they could speak in the Commons" (O'Neill and Dalrymple 2008: A12).

Despite accusations of government hypocrisy and duplicity, the occasion of the unprecedented and overdue apology was deeply moving. On 11 June 2008, from our home in Nanaimo, British Columbia, my husband, Pete Steffens, and I wept as we watched the coverage of the day's events on CBC, Canada's public broadcaster, and on APTN, the Aboriginal Peoples Television Network. The live broadcast and commentary on APTN was itself a testament to the survival of first peoples. APTN provided full, live coverage, and streaming video available 24/7 on its website. Throughout the proceedings, its reporters kept reminding viewers of the AFN's 24-hour phone line for survivors and others needing support. Holly Bernier was anchor and host, aided by APTN Investigative Journalist Paul Barnsley (Bernier and Barnsley 2008).

CHAPTER 4

Turning the Camera and Microphone on Oneself

Pirates and Precedents

Earlier, I called The New Media Nation an "outlaw" nation. According to *Webster's Online Dictionary* (2008), the English noun, outlaw, dates back to before the twelfth century, evolving from the Old Norse *ūtlagi,* from *ūt* (out) + *lag- or loāg* (law), to the Old English *ūtlaga*, Middle English *outlawe*. The current meanings are: "a person excluded from the benefit or protection of the law"; "a lawless person or a fugitive from the law"; "a person or organization under a ban or restriction"; "one that is unconventional or rebellious"; and "an animal (as a horse) that is wild and unmanageable." The *Compact Oxford English Dictionary Online* (2008) places the two main definitions in reverse order. Here, the first definition is "A fugitive from the law," and the second, "A person deprived of the benefit and protection of the law." I would contend that all of these definitions, including the one about "unmanageable" animals, can be applied to the spirit and substance of the New Media Nation, with the "unmanageable" label falling into the (racist) "wild Indian" genre of outsider misrepresentation. Most of the Indigenous media practitioners and consumers I have interviewed, over the years, view the "Nation's" outlaws as Robin Hood figures, operating in the people's interests. If not literally "stealing from the rich," they most certainly are "giving to the poor"—to those disadvantaged by governments, corporations, and more generally, by economic and environmental circumstances.

Groups such as the Ainu Rebels express "outlaw" activism in their very names. In some places, "outlaw" or "guerilla" media are a necessary part of survival, both for the information they provide, and for the protected, sometimes secretive nature of their operations. Lisa Brooten observes that the "Thai-Burma borderland has for decades provided

Burmese activists with the freedom to develop and distribute alternatives to the highly censored media in Burma." (Brooten 2008: 115) "Numerous clandestine radio stations have operated from these border areas" since Karen people started *Radio Kawthoolei* in 1949. "Both armed and nonviolent resistance groups inside Burma have used wireless transmitters for local radio communications", and journalists with international broadcasting networks also operate from the border region (Brooten 2009: 116). Throughout history, Indigenous media projects have often begun in "illegal," "outlaw," "guerilla," "rebel," or "pirate" ways. In the northern region of the Canadian province of Ontario, the first radio broadcasts were carried through the countryside by equipment installed in a traveling van. Various versions of mobile services have emerged worldwide throughout the history of Indigenous broadcasting. In Suva, Fiji, Femtalk 89.2, the women"s community radio initiative of femLINKpacific (Media Initiatives for Women), uses a mobile "suitcase" radio to broadcast to women in semi-urban and rural Fiji (Mutuku 2005). In the late 1980s in Salt River, Cape Town, South Africa, a small group of people founded CASET (Cassette Education Trust) to produce radio-formatted cassettes for distribution in townships and in the area around Cape Town, "to inform and educate impoverished people on issues such as literacy, hygiene, health, and politics" based on a philosophy of "Information is POWER." In 1992, CASET transformed itself into Bush Radio, a "voluntary association owned and operated by its members." According to CASET, this was "the first time in the history of South Africa [that] 'black' people ... have the opportunity to be broadcasters" (Mott Foundation 2005).

In 2007, the US-Senegalese film production company, Nomadic Wax & Sol Productions, released its feature-length documentary, *African Underground: Democracy in Dakar*. Filmed in Wolof and French and screened with English subtitles, it describes the courageous and influential participation of hip-hop artists in Senegal's political process. In a climate of intimidation and fear, when many people were afraid to speak out, Senegalese hip-hop artists continued to perform their message-laden music and to advocate and promote the democratic process. This 'legitimate,' publicly screened film portraying activist popular culture, was the child of outlaw media. "Originally shot as a series of shorts distributed via the Internet, *African Underground: Democracy in Dakar* explores the boundaries of guerilla-style film production and distribution (Africa-in-motion 2009)."

When the telegraph arrived, in the 1840s, communication was no longer inherently linked to transportation. This leap, from what is called a "transportation" model of communication to a "transmission one," was not without its precursors. Talking drums, smoke signals, and the use of polished metal to direct sunlight (heliograph) were early ways of sending messages without messengers (Crowley and Hyde 1991: 145).

In remote communities, news media often begin as "alternative" services. In Arctic Norway and Finland, Sámi radio services sometimes arrive by snowmobile. In northern Ontario, the Wawatay Communications Society started as a trail radio rental service to trappers, with small, high-frequency transmitters forming an emergency communications system. The evolution of Indigenous media in Canada can be traced to the moment when the Métis leader, Louis Riel, seized Fort Garry, imprisoned John Christian Schultz and other media and political opponents, took over the printing establishment, and began publishing the *New Nation.* During the 1880s, newspapers emerged throughout the West, along with the new towns that accompanied the building of the Canadian Pacific Railway. The link between transportation and communication was underscored by publication of the *Medicine Hat News* from an aging Canadian Pacific Railway boxcar. That tradition continued in the 1990s, when for several years Miles Morrisseau and Shelley Bressette published *Nativebeat* newspaper from a trailer at Kettle Point, Ontario.

Figure 4.1. Shelley Bressette (L) and Miles Morrisseau, founders of *Nativebeat,* with colleagues in the office next to their home, Kettle Point First Nation, Ontario, Canada. Photograph by Valerie Alia.

Figure 4.2. Outside view of the *Nativebeat* office, with Miles Morrisseau at the door. Photograph by Valerie Alia.

In Yukon, the Klondike gold rush brought new cities and newspapers to the North. Northern broadcasting had its origins in early radio broadcasts that were often closely tied to newspapers, setting the stage for today's media monopolies and common ownership of print and broadcast media. In 1917, the Canadian Press (CP) wire service ended railway control over transmission of the news, and in 1923, CP's eastern and western wire services were merged into a nationwide service. During the Winnipeg General Strike of 1919, The *Western Labour Press* published a strike edition and the *Winnipeg Free Press* an anti-strike edition. Finding itself cut off from the wire services, the *Free Press* installed a transmitter atop its building and began transmitting radio broadcasts. In the next decade, commercial radio stations began to emerge across the country. Echoing Winnipeg, the *Calgary Herald* newspaper installed a major broadcasting station atop its building, and heralded the contemporary era of communications, with its blending of print and broadcast boundaries.

In First Peoples' communities, (especially in remote areas) meetings, feasts, potlatches, and public events of all kinds effectively become communications media—much as early troubadours and town criers brought the news to street corners, coffee houses, and pubs. In Indigenous communities, services often begin as underground or aboveground "alternative" media. As in the case of the Wawatay Communications Society and the *Nativebeat* newspaper, small-scale services are based in, and responsive to, their communities.

"Alternative" media merit full attention and credibility. In fact, the very term "alternative media" is misleading and is based on colonialist notions of "core" and "periphery." Especially where Indigenous communications are concerned, "alternative" does not mean "unprofessional" or "inferior." Innovative survival strategies start as "alternatives" and evolve into so-called "mainstream" communications. Pirate stations go "legit," bulletin boards become newsletters, newsletters become newspapers or magazines, local newspapers or magazines become regional, national and sometimes international.

Life on the Margins

As we will see again in the discussion of the African bush radio experience, "alternative" and non-legal media occupy a significant, though sometimes transitory, place throughout the Indigenous world. In Canada, Lorna Roth notes that the "history of appropriating airwaves … has tended to be related to the illegal capture and use of an airwave frequency" and tells the story of the beginnings of Inuit-run, Inuktitut-language radio. In 1964, a group of Inuit "innocently put together an inexpensive radio station and began to broadcast culturally-relevant programming without going through the conventional regulatory channels." Several years later, two pilots flying over Montreal's Dorval airport heard Inuktitut on the airwaves and were baffled by the mysterious language.

> Thinking it was Russian … they had the air traffic control tower employees research the source [and] discovered that the voices originated from the tiny community radio station in Pond Inlet. Apparently, the sound waves had traveled unusual distances due to atmospheric abnormalities on that particular day. The radio station was contacted … and procedures were normalized accordance with Canadian broadcasting regulations (Roth 1993 [unpaginated]).

Given the history, it is probably no coincidence that Indigenous media in Canada developed first in the Arctic and sub-Arctic regions. Abraham Tagalik, Television Northern Canada (TVNC) Chair and Director of Network Programming for the Inuit Broadcasting Corporation (IBC) in Iqaluit, was the first chair of APTN, Canada's nationwide network.

While strongly advocating for Indigenous media, Miles Morrisseau has expressed his concern about the "dangerous control Aboriginal organizations have over the [Aboriginal] media" (Morrisseau 1998). His concern about too much government control (whether by Indigenous or other governments) does not mean—as some have suggested—that simply "going private" will solve the problem. As the journalist, Helen Fallding, said during her time at CHON-FM, Northern Native Broadcasting Yukon, advertisers eliminate the problem of government-run media, but bring new concerns about their own efforts to control content and presentation, and sometimes even the journalists themselves. She said,

> I have lots of concerns about the move to commercial aboriginal radio. I have no idea how it's all playing itself out at this point at CHON-FM, but while I was there, there was more and more white business "expertise" at the management and advertising sales level, aimed at increasing the station's commercial viability. The people hired had no clue about aboriginal culture or values and didn't seem interested in learning anything. The result [included] some terrible decisions. One top administrator saw the station as essentially a country-and-western station whose target audience should be C and W [Country and Western music] listeners, not Aboriginal people. An ad salesperson designed ads for a strip show at a local tavern, which were aired. Perhaps these were growing pains (Fallding 1995).

A Kaleidoscope of Media Projects

The International Initiative for Community Multimedia Centers (CMCs) crosses the digital divide through a combination of community broadcasting, Internet, and related technologies. A Community Multimedia Center brings together community radio broadcast in local languages and telecenters providing public access to internet-linked computers, phone, fax, and photocopying services. The services help develop community databases, education, and training (UNESCO, 10 February 2006). UNESCO and other private and public agencies have funded a number of training and production initiatives in Indigenous radio, television, and new media. The Indigenous People and Information Society program fostered a series of pilot projects in 2004/05 and 2006/07. Some examples follow.

Forest communities of Gabon

Bakas, Babongo, Akula, and Bakoya—often called "Pygmy"—were involved in a project directed by French filmmaker-producer Jean Claude Cheyssial and Mynapiga, a Gabonese NGO. In 15 days, six participants produced four 13-minute documentaries, which were assembled into a 26-minute film. They also produced a music video of Rasta singer, Moses, and a short documentary on Bakoya snail hunting. The films were screened in traveling cinemas, at cultural centers in neighboring countries, and on several television stations. Responding to their experience, two of the participants wrote:

> *We are Bakoyas, we live in the forest... We live from hunting and fishing Pygmies are also men like all the others They treat us bad, they bully us all the time We are Pygmies, we are Bakoyas.*
> —Estimé Mbithé

> *Today I have been trained to use the camera, tomorrow I will become a cinema man.*
> —Jean-Louis Ngoubamé

Native Network, Bolivia

Bolivia has a population of about eight million. Although about 70 percent are Indigenous, with 36 distinct peoples (tribes), they are "the most vulnerable and marginalized sector of society. The National Plan aims to strengthen Native self-representation by appropriating technology to create indigenous peoples' own images." Using the long tradition of community radio, the National Plan started by creating National Network of Communication and Audio-Visual Exchange in more than 100 of Bolivia's Indigenous communities. In the first five years, it produced about 70 works, including video series of "Indigenous fictions," some of which were featured in the NMAI Native American Film and Video Festival 2000.

Audiovisual and Community Television Production in Bolivia

Run by the "Centro de Formación y Realización Cinematográfica" (CEFREC), whose director is Bolivian filmmaker Iván Sanjinés, the project took place in the districts of La Paz and Beni in northwestern Bolivia. They included several of the region's Indigenous communities: the Lecos, Tsimanes, Esse Ejjas, Mosetenes, Tacanas, Quechuas, and Aymaras. The project was linked to the "Plan Nacional Indígena de Comunicación Audiovisual" (Bolivian National Plan for Indigenous Audiovisual Communication), inaugurated in 1997 following lobbying by Coordinadora Audiovisual Indigena Originaria de Bolivia and CEFREC. Twenty-three Indigenous leaders were chosen, with advice from local organizations and authorities. In a screenwriting workshop, each trainee produced a script, and five of the scripts were selected for production. The participants produced four 25-minute documentaries and a fiction film. These were shown at community screenings and on regional and national television. The project also assisted program development in the Indigenous community television station at Alto Beni. Participants said:

> *As a woman it has been very significant to get trained in audiovisual editing ... I have been able to propose the idea of a documentary to my community.*
> —Maria Morales (Tajlihui-Larecaja community)

> *The training center in Sapecho is a strategic place ... a great step forward for the region as it is run by local people who care about ... culture and social issues and try to improve the well-being of their community.*
> —Esteban Espejo (Piedras blancas-Rurrenabaque Community)

> *We are doing this work so that the region develops its own expression, because through this way we can reflect the problems and the needs of our communities.*
> —Jesús Tapia (Belén-Alto Beni Community)

The San Interactive Archive Training and Heritage Management Program, South Africa

This project involving San people was run by the Khwa Ttu San Centre and Mindfield LAMP (Living Archive Management Project) in collaboration with the Working Group of Indigenous Minorities in South Africa (WIMSA) and Doxa Productions. Participants selected by their communities received training in audiovisual techniques of cultural curating, including audio-visual archiving systems. Participants produced a short video on the use of language in memory retrieval and cultural practice and developed a pilot for a multimedia interactive archival package that will be used for producing documentaries, short narratives, animations, and other educational materials.

The Tokapu Video Workshop in the Villa El Salvador District of Lima, Peru

The Tokapu project was organized by the "Asociación Cultural Integración Ayllu-Wari," along with French filmmaker and anthropologist Elif Karakartal. Villa El Salvador is a district in Lima inhabited principally by Quechuas, who came there from the Sierra starting in the 1940s. The participants were second-generation Quechuas, who had grown up in a multicultural environment. The objective was to initiate an intercultural dialogue between "westernized" and "traditional" Quechuas, and between Indigenous and non-Indigenous people from diverse ethnic and cultural backgrounds. Participants produced ten 20-minute video documentaries. The films were shown in a series of public screenings. One Quechua participant said that he now "would like to shoot a documentary about the problems of my community."

The Igar Yala Collective: Tradition and Innovation in Panama

Here is an example of the marriage of cultural preservation and renewal with innovative mediamaking and change. Founded in 2004 by writer/director Vero Bollow, the Igar Yala Collective is a group of filmmakers and artists who are dedicated to preserving cultural memory, collective/collaborative writing and filming, and to using arts and media for community empowerment. In the process, they are discovering and creating new ways of making film. Most of the collective's members are young rural and urban Kuna adults. Working with international artists, in collaboration with Kuna elders, they bring their own stories and their community's cultural traditions and experiences to the screen. The collective's first major project was the 2008 feature film, *The Wind and the Water* (detailed in the Filmography).

The Kaoko Local Knowledge Living Archive Project: The Well of Stories

The project aims to "map the memory of the Himba people ... and contribute to the continuance of a way of life that is constantly challenged." In the Kaoko region of northwest Namibia, it involves production of an interactive DVD-Rom as a living archive of the Himba oral culture. The pilot DVD includes a film on the life of oral historian Katjira Muniombara, ICT training of one member of the Himba community, and preliminary research toward creating a multipurpose ICT community center in a Himba village. "Living Archive" packages will be accessible in Herero, the Himba language, and to others via the Internet and on DVD.

Lao People's Democratic Republic

Ta-Oi Radio began broadcasting in January 2006, becoming the first regional station in the Lao People's Democratic Republic. It serves people from several different ethnic groups, in the remote Ta-Oi district of Saraven Province, located 700 kilmeters south of the capital city of Vientiane. "The Provincial Agriculture section contributed a part of its centre to house the radio studio," which has a staff of five and produces about six hours of morning and evening programming a day. Staff training was provided by the Mass Media Department of the Laos Ministry of Information and Culture and the Laos National Radio. Programming focuses on "education, health and agriculture with special attention to the needs of the Ta-Oi ethnic groups" (UNESCO 2 July 2006).

From Bush Radio to Blogging in South Africa

"The news of South Africa's dramatic pledge to transform apartheid reached many of its citizens via radio. A decade later, that medium remains critical to many of South Africa's remote communities. The historic elections of 1994 led to the transformation of many South African institutions, including its media, noted John van Zyl, managing director of ABC Ulwazi." Government deregulation and establishment of community radio has meant that more than 90 percent of the country's rural population tunes in to listen. Russell Ally, Director of Mott's South Africa office, thinks community radio is an essential "watchdog over civil liberties" and a "vehicle for local civic participation." ABC Ulwazi has a network of some 60 community stations, which collectively produce more than 2,000 hours of programming a year. ABC Ulwazi plans to make the service more interactive. To help strengthen citizen participation in radio programming, the Mott Foundation helped to fund listeners' associations in ten communities, based on the view that listeners' associations provide "the link between the radio station and the community" (Mott Foundation 2005).

"In African countries where newspapers are state-owned or censored [people turn] to the blogs." "Blogs are taking off across Africa as a new tech-savvy generation takes advantage of growing internet access. The African blogosphere was, until recently, filled by the African Diaspora and westerners living in Africa. But native African voices are now being heard." The Kenyan Blogs Webring started in 2004 with ten sites. In 2007, there were more than 430. "In parts of Africa where the media is tightly controlled, blogs have emerged as an essential tool in highlighting injustices." Sometimes they can only be read outside the country. For example, Ethiopian government blocks anti-government blogs. "Much of the best on-the-ground reporting from Darfur has been done by bloggers rather than journalists" (Mott Foundation 2005). Blogging can contribute to community-building in the best, and the worst, sense. It can cross geographic, cultural, and political borders in a mutually supportive way, or it can reinforce chauvinistic, nationalist insularity.

CASET pressured the South African government for a license to broadcast, develop a nationwide training program, and to internationally transmit and promote its broadcasts. Frustrated by several rejections from the government, Bush Radio began broadcasting illegally, in May 1993, when about 20 volunteer activists took over the station's 16-channel studio mixing desk and illegal transmitter, and a supply of CDs and tapes, and began broadcasting. The inaugural four-hour broadcast was the last for several months. Authorities invaded the premises and seized the equipment, and charged Bush Radio's two key members on counts of "illegal broadcasting, illegal possession of broadcast apparatus, and obstructing the course of justice." After a huge public outcry and eight months of pressure from individuals and from national and international organizations, the state dropped the charges (Mott Foundation 2005).

On the eve of the country's first democratic elections, Bush Radio began its broadcasts and launched its national training program. It was the first South African community radio initiative to join the World Assembly of Community Radio (AMARC), and has continued to lobby and help develop community radio across South Africa. The training program is premised on the decentralized, widely distributed training of representatives of each community station who return to their own stations and train the volunteers (Mott Foundation 2005). Shortly after it was founded in 1994, the National Community Radio Forum (NCRF) undertook more than a 100 license applications that were to rapidly and radically expand the South African broadcasting universe. In June 1995, seven years after CASET began its campaign to "get the people on air," Bush Radio received its license. Today, it has partnerships and exchange programs in several countries, an Institute for the Advancement of Journalism and Media Training Center, and a national resource for radio production and management skills.

Tales, Sales, and Cells: "Guerilla" Phones in Africa

In several African countries, mobile (cell) phones are fostering economic and social change. "Mobile reporters" send news stories to and from "all corners of Africa," with the help of a cooperative project between the Dutch mobile reporting portal, skoeps.com, and the *Africa Interactive Media Foundation.* Citizen-journalists use the phones to write articles, film interviews, and disseminate images and text. One 2009 report, which the mainstream media missed, publicized the abysmal water quality in Kenyan slums (Kreutz 2009).

Starting in Kenya, Africans "have pioneered the use of cellphones to transfer value by using airtime as a virtual currency." In Zimbabwe, Mukuru.com facilitates mobile phone money transfers, such as payments for gasoline (petrol). "For instance, gas fuelling can be paid over the Internet from anywhere to anybody with a mobile phone in Zimbabwe, then the petrol station owner gets his money back through vouchers" (Kreutz 2009).

In Rwanda, mobile phones support HIV care, not just by informing the public, but by improving health service itself. Healthcare facilities often lack supplies, reliable Internet connections, and ways of tracking patients and the spread of illness. A service called *Phones for Health* allows health workers in the field to transmit data to central computer systems, improving the ability of health officials and service providers to respond. By 2009, after two years in operation, *Phones for Health* had linked 75 percent of Rwanda's 340 clinics and 32,000 patients, and is now being extended to several more countries (Kreutz 2009).

A mobile phone network called TradeNet operates in four languages and informs sellers and buyers about pricing and crops, in real-time, sourced from 380 markets across Africa (Kreutz 2009). Alan Brewis, an engineer, farmer, inventor and manager who has worked for many years with NGOs in Zimbabwe and Ghana, observes that cell phones are in widespread use throughout Africa, especially by farmers and fishers. By communicating directly with potential buyers, they

are able to increase their profits through direct sales of crops and fish, avoiding the "middle men" (Brewis 2009). This has had particular impact on women. According to a three-year pilot study by the Regional Hunger and Vulnerability Programme, access to mobile phones "transformed the lives" of women farmers in Lesotho, increasing their income and providing valuable new information. Ten mobile phones were distributed to three cooperative women's farming groups in different agro-ecological zones in Maseru district, western Lesotho. The phones have improved the effectiveness of marketing produce, accessing and exchanging information on prices, and more generally, enhanced self-confidence, according to Gladys Faku, national chairman of the Participatory Ecological Land Use Management (PELUM), a network of NGOs and civil society groups working with small-scale farmers in East, Central and Southern Africa (IRIN 2009).

The phones have "reduced the time and cost of staying in touch." The women have also used them as an income-generating tool. The women sell airtime on their phones, and use that income to purchase more phones, thereby extending their mobile network. One group also used the money to buy piglets, which were sold to generate more money" (IRIN 2009).

In Kenya, cell phones are used to transfer cash to vulnerable people. The intention was that the recipients would use about one-quarter of this for group communication and then sell the remaining amount as airtime to other community members, so that the enterprise would become self-sustaining. A follow-up evaluation in January 2009 found that the biggest saving was in time and travel costs in mountainous Lesotho, which has enormous distances and a poor public transport system. Among other benefits are shorter waits at health centres. "The women phone in advance to get people to queue for them," said Vincent.

Michael Posluns' observation about the limits of cell phone use among Indigenous people in Canada is also true of Africa: the phones and airtime are expensive and beyond the reach of most people. Those used for the Lesotho project were donated by Vodacom Lesotho, and a certain degree of caution is called for, in terms of corporate motivation to expand their own markets. However, the phones are gradually becoming less expensive, and some charitable organizations are providing people with recycled phones. Most communities lack access to electricity and many send their phones to town for recharging. RHVP optimistically recommends the use of solar powered chargers, but the reality is that they are often too costly.

The Guatemala Radio Project

"Guatemala's Indigenous peoples have faced more than 500 years of persecution and decades of civil war. In the last war, more than 200,000 indigenous people were murdered" (Cultural Survival 2005). Although they are the numerical majority in Guatemala, Indigenous peoples "have very little representation and voice in the government." Where governments fail to provide access or adequate funding, NGOs and private funding can help to close the gap. In 2005, Cultural Survival started the Guatemala Radio Project, to support local stations broadcasting to Maya peoples across Guatemala. This important project could be open to compromise; its sources of support include members of the US Congress and the Department of State. Cultural Survival brought "selected representatives" of Guatemalan communities to the US to meet their benefactors and promote their cause. Their radio interviews included the Voice of America. Cultural Survival's annual report called it "the first international media attention for the Guatemalan community radio movement." (Cultural Survival 2007: 5)

Such attention can be a mixed blessing. This is an extremely important service and a cultural and linguistic lifeline, which helps to strengthen, preserve, and perpetuate Indigenous languages. We must also ask whether what Guatemalan peoples are allowed to say in those languages is restricted. While Cultural Survival has a strong record of providing support without strings or outside agendas, the US interests may not be entirely altruistic. Somewhere, there are links to people investing, or wishing to invest, in Guatemala. Who initiates, prioritizes, regulates, and controls those investments is an important part of the picture. I asked the historian and journalist, Pete Steffens, whether such vested interests negate the project's value. He thinks that that is not the case: "History is dynamic. The more things are moving, the better. Regardless of the motives of different benefactors, the fact is that people are getting power and their voices are being heard" (Steffens 2007).

Commercial radio and television stations broadcast only in Spanish. Even if TV sets were affordable, television signals "rarely reach remote communities." Community radio stations "broadcast in all of Guatemala's indigenous languages and the abundance of small battery-powered radios makes their content accessible to even the most isolated communities." (Cultural Survival 2005) The initial stage was staffed and programmed by Cultural Survival, who worked with Guatemala's five Indigenous radio associations and Guatemalan NGO partners, to train local journalists and technicians The plan is to obtain private funding to properly equip the stations, so that Indigenous people can produce and disseminate their own programming, and to use this as a model for developing projects in other countries.

Local community radio stations "promote indigenous music, language, and culture," and broadcast information in Indigenous languages. Cesar Gomez, a radio volunteer, said, "Before we started the radio station in Palin Escuintia eight years ago, our language, Pocomam, was only spoken in our homes. Now Pocomam is spoken everywhere—in offices, in the streets, even by our children, who are learning it from their parents. Without community radio, we might have lost our native tongue" (Cultural Survival 2006: 5).

The theme of Cultural Survival's 2006 appeal campaign was, "Democracy with a microphone." Beneath a photograph of Elena Yach holding a microphone, is the text: "Because she listened to a Guatemala Radio Project broadcast in Nahualá, Elena Yach and her friends found out they had the right to wear Mayan dress to school. They told their teacher, and she agreed to change the dress code … In Guatemala, 250 community radio stations are broadcasting in 15 languages to seven million Maya who have no other reliable source of information" (Cultural Survival 2006, unpaginated).

The Guatemalan government officially recognized the importance of communications media in one of the provisions of a peace accord that ended a decade-long civil war. Previously. there was no legal provision for telecommunication services and the government could not "officially assign bandwidth to the radio stations." While the accord paves the way for new telecommunications legislation, a full solution remains elusive, especially in light of more recent government backsliding.

According to Mark Camp, director of operations for the Cultural Survival project, "Guatemala is one of only two Western Hemisphere countries in which indigenous people are a majority of the population. It also has a 500-year history of violence, with its civil war ending only 10 years ago." He sees telecommunications as crucial to informing people and encouraging them to participate in the country's political life. "Not everybody can read, but everybody can listen to a radio." (Cherrington 2006: 14)

Spanish-language commercial radio and television leaves Maya peoples out of the picture. Television signals cannot reach most remote communities, and even where they do, TV is generally unaffordable. With a high rate of illiteracy, print media are not widely accessible. Community radio stations broadcast to Maya peoples throughout Guatemala and "play an important role in fostering democracy and inter-ethnic tolerance," by filling a significant media gap. All of Guatemala's Indigenous languages can be heard on community radio, "and the abundance of small battery-powered radios makes their content accessible to even the most isolated communities" (Portalewska 2005).

While framed somewhat paternalistically, the program continues the long established policy of making communications services available to people in all parts of Guatemala, and provides a foundation for the next stage of communications, the continuing development of telecommunications serving Indigenous people. The program's vulnerability was demonstrated once again in July 2008, when Guatemala's Public Ministry reverted to the practice of raiding community radio stations and seizing their equipment. Four stations in the department of San Marcos were raided. The raids go against both the Peace Accords and the Guatemalan Constitution, which "guarantee indigenous Mayans the freedom to use community media." Failure to update the country's telecommunication law means that any "'man of influence' can have the police raid a station, arrest the operators, and confiscate equipment" if the station is broadcasting without a license. The "Catch 22" is that under present law, the stations cannot be licensed. The process of legislating "to regularize the 140 community stations" has been slow, and the confiscation of equipment leaves thousands of Indigenous Guatemalans without "their only source of vital news and information" in addition to their financial losses (Cultural Survival September 2008 [online]).

Media Projects in Mexico

In their study of Indigenous Mexican media developments, Amalia Córdova and Gabriela Zamorano quote a statement by the Tzeltal (Indigenous) videomaker, Mariano Estrada Aguilar, which could easily be transferred to any number of Indigenous media projects:

> In reality I am not an independent videomaker—while the technical questions of videomaking are solved individually, the feeling and content of my videos belong to the people (Córdova and Zamorano 2004: 1).

Indigenous media became active in Mexico in the 1980s. One of the earliest video efforts was a 1985 video workshop with weavers, held in San Mateo del Mar, Oaxaca, as part of a documentary series created for the Instituto Nacional Indigenista, now called the Comision Nacional para el Desarollo de los Pueblos Indigenas de México (CDI). Since the mid 1990s, Indigenous videomakers have been active throughout Mexico. There are now more than a dozen independent Indigenous media organizations and four regional Indigenous media centers (CVIs) that are supported by government funding. The authors of the report note "an ongoing conversation among communities, media organizations, independent producers and audiences ... with increasing interregional collaboration. Video Mexico Indígena/Video Native Mexico (VMI) has sent programming to the US, and the Chiapas Media Project-Promedios (CMP) took works of Mexican Indigenous videomakers on tour to Australia, in collaboration with Australian Indigenous videomakers" (Córdova and Zamorano 2004: 2, 5).

In 1992, Indigenous videomakers formed the Organización Mexicana de Videoastas Indigenas (OMVIAC) with a goal of creating a national organization. Today, there is a cluster of independent Indigenous media production centers and eight state-sponsored Indigenous Audiovisual Production Units (UPAI). Major projects throughout the country include:

Oaxaca

Transferencia de Medios Audiovisuales, with Zapotec and Mixtec directors; the independent media center, Ojo de Agua Comunication; Solidario de Quiatoni in the Zapotec region of the Sierra Sur; the Centro del los Derechos de la Mujer Naawiin that focuses on women's rights; and the regional Unión de Comunidades Indigenas de la Zona Norte del Istmo (Cordova and Zamorano 2004: 3).

Chiapas

The Chiapas Media Project has produced a number of videos, including *The Land Belongs to Those Who Work It,* which documents a dispute between resort developers and local communities. "We're not just documenting resistance … Chiapas Media Project is part of it; a form of resistance in itself," Alexandra Halkin of the Chiapas Media Project has said. She explains that the absence of mainstream, national news coverage of remote, isolated regions often makes CMP "the *de facto* protector of law and order for Indigenous communities. For example, CMP videos of military incursions into Chiapas, where government soldiers [were] caught throwing rocks at local children, have been used successfully in court, forcing the government to pay the children's hospital bills" (Verán 2006).

Figure 4.3. The Chiapas Media Project documented this 2003 march by Zapatista activists in Oventic, Mexico, celebrating the proclamation of regional autonomy for the indigenous peoples of Chiapas. Photograph by Francisco Vazquez for Fellowship of Reconciliation.

In the Lacandon communities, since 1995, "Casas de la Cultura" (houses of culture) community centers have been developing as part of a broader program to preserve Lacandon culture and identity. The aim is to provide hardware and workshops "to make the world wide web accessible to communities in rural, south eastern Chiapas." The challenges to this project are daunting. The climate is hostile to electronics, there are no telephone lines, and computerized information systems are unknown (Hach Winik Home Page 2006). The situation, like so many others, calls for technological acumen and human brilliance, persistence, and fortitude.

Guerrero

Promedios de Comunicación Comunitaria is active here, along with local organizations and the community media organization, Altepetl Nahuas de la Montaña de Guerrero, and the independent production company, Ojo de Tigre Comunicación/Mirada India (Córdova and Zamorano 2004: 4).

Michoacán

Centro de Video Indígena was launched in 1996 and is very active, with youth video workshops, traditional and contemporary P'urépecha music, and television and radio programming (Córdova and Zamorano 2004: 4).

Yucatán

Yoochel Kaaj/Cine Video Cultura works in Mayan communities, while others work with participants from Maya, Zapotec, Mixtec, Tzeltal, and Chol communities (Córdova and Zamorano 2004: 4).

Sonora

The city of Hermosillo has a CVI serving Yaqui, Mao, and Seri communities of northwest Mexico and an Indigenous Audiovisual Production Unit, UPAI, working in Sonora (Córdova and Zamorano 2004: 5).

Women's Listening Clubs in Zimbabwe

In Zimbabwe, the Development through Radio (DTR) project has fostered 52 women's radio listening clubs, which give rural women access to radio through participation in program production (Mufune 2004). DTR uses radio-internet convergence in Sierra Leone. Women's listening clubs' programs are digitized and archived (Development Gateway 2003).

Solar and Satellite Power in Niger

In Baniklare, a region in Niger populated by nomadic people, solar-powered WorldSpace satellite receivers and Baygen Freeplay windup/solar powered FM radios deliver programming from the WorldSpace Afristar satellite (Benamrane 2000; Hijab 2001).

Low-tech, High Achievements: the International Programme for the Development of Communication (IPDC)

UNESCO supports a cable television project in Bangladesh with the Tripura adivasi people, who have lived in the same region for several centuries. Literacy is low and there was formerly no access to television or computers. In 2006, community members asked a local NGO, Youth Power in Social Action, to provide them with a television, and the Chairman of the Sitakund Municipality was asked to provide a Video Compact Disc (VCD) player, and they organized a community fundraising project to buy a diesel generator. Each week, they borrow local programs produced by volunteers of CMC Sitakund and view them in the community school. In partnership with Indigenous communities and with support from UNESCO, YCMC is developing a series of audio-visual documentaries. They say that cable broadcasts and narrowcasts of the documentaries will help to inform the wider public. "We migrated to Sitakund hundreds of years ago and have since adjusted to the local Bengali culture. Only a handful of the elderly people in my village now remember songs in our language," says Lakshmi Tripura. These were the reactions that staff of the Youth Community Multimedia Centre (YCMC) Sitakund encountered when they visited the village of the indigenous community, Choto Kumira Tripura Para, [in] Bangladesh. "'We have no wealth, nor power or respect, what was the use of taking birth in this world'—this was the literal translation of the indigenous song performed by an elderly singer—one of the last remaining indigenous artist of Sitakund. Now we need to see if new media technologies can play a role in transforming this situation and how." The YCMC uses the local cable network [to reach] about 1200 households (UNESCO 2006).

New Alliances in the United States: United South and Eastern Tribes, Inc. (USET)

On 4 October 1968, the Eastern Band of Cherokees, the Mississippi Band of Choctaws, the Miccosukee Tribe, and the Seminole Tribe of Florida met in Cherokee, North Carolina, "with the shared idea that some form of unity between the Tribes would facilitate their dealings with the federal government." A year later, they formed the inter-tribal council United Southeastern Tribes, based at Emory University in Atlanta, Georgia, later moving to Sarasota, Florida, before ending at their current headquarters in Nashville, Tennessee. The organization changed its named in 1978 to United South and Eastern Tribes, Inc. (USET) to better reflect its membership, as federally recognized Tribes from Maine, to Florida, to Texas affiliated themselves with the organization. Today, USET comprises 24 federally recognized tribes who collaborate through workgroups and committees. USET's current membership includes the Eastern Band of Cherokee, the Mississippi Band of Choctaw, the Miccosukee Tribe, the Seminole Tribe of Florida, the Chitimacha Tribe of Louisiana, the Seneca Nation of Indians, the Coushatta Tribe of Louisiana, the St. Regis Band of Mohawk Indians, the Penobscot Indian Nation and Passamaquoddy Tribes, the Houlton Band of Maliseet Indians, the Tunica-Biloxi Indians of Louisiana, the Poarch Band of Creek Indians, the Narragansett Indian Tribe, the Mashantucket Pequot Tribe, the Wampanoag Tribe of Gay Head (Aquinnah), the Alabama-Coushatta Tribe of Texas, the Oneida Indian Nation, the Aroostook Band of Micmac Indians, the Catawba Indian Nation, the Jena Band of Choctaw Indians, the Mohegan Tribe of Connecticut, and the Cayuga Nation (USET website).

Crossing Borders in Sápmi

In the multicountry region of Sápmi, the Sámi parliaments are divided, country-by-country, because of the policies of different governments. Overall, Sweden is the least progressive of the Sámi countries. Yet many projects—especially media projects—cross the artificially created national borders and transcend Sámi relations with the separate states. Norwegian Sámi radio and television collaborate with Finnish, Swedish, and Russian broadcasters, while at the same time each must deal separately with the policies, funding, and restrictions of its home government.

Much as John Amagoalik sees the strengthening of Inuktitut (Inuit language) broadcasts and education as integral to survival, Katrine

Johnsen, known as "the Mother of Sámi Radio," saw language as crucial. Ivar Bjørklund has said that she "gave the Sami language a place in the media" (Bjørklund 2000: 21). Based in Norway, she was heard on Norwegian radio for almost thirty years, from 1959 to 1987. Her legacy continues across Sápmi, where Sámi language broadcasts continue, and increase.

Norwegian, Swedish, and Finnish national broadcasting companies have an agreement to share television newscasts. They started small, with a daily 15-minute Sámi-language newscast accompanied by Norwegian, Swedish, and Finnish subtitles. "Our surveys say that we have about 70.000 viewers every day all over the year," confirms Nils Johan Heatta, the director and editor-in-chief of *Sámi Radio* (Arbelaitz 2007). Lander Arbelaitz, a Basque journalist looking at Sámi media, reports: "Without making much noise, [Sámi] are building up their culture and gaining the rights, which belonged to them long time ago. Some Sámi call this a 'silent revolution'" (Arbelaitz 2007).

In Norway, Sámi Radio, comprising radio and television services, is a division of the Norwegian Broadcasting Corporation (NRK). Its Director, Nils Johan Heatta (spelled Haetta by non-Sámi), reports directly to the Director General of NRK, and is responsible for all Sámi radio and television broadcasting. As a public service broadcaster, NRK is committed to promoting "Sámi language, culture and identity" nationally,

Figure 4.4. Exterior of the Sámi Radio headquarters at Karasjok, Norway. Photograph by Valerie Alia.

Figure 4.5. Nils Johan Heatta, Director of Sámi Radio, Norway, in his office at Karasjok. Photograph by Valerie Alia.

regionally, and locally. "As such Sámi Radio—within the Norwegian Broadcasting System—should stand up as an innovative, versatile, reliable, indispensable and well-organised enterprise" (NRK Sámi Radio website 2006).

There are two categories of programs—regular Sámi programs and those made for other departments. Regular programs include a 15-minute children's TV show broadcast every other week and a monthly half-hour documentary, feature, and debate. There are also independent Sámi-produced documentaries about Sámi and other Indigenous peoples. Programs made for other departments consist of Sámi-language news reports, short features for youth and adults, programs made by NRK's regional departments, and environmental documentaries.

As with NRK, the Sámi divisions are financed by license fees. "Since 1984, Sámi Radio has its own broadcasting station in Karasjok. Located near legend-shrouded Stálloláttu, the station has been extended twice, comprising…offices, editorial premises and production facilities for both radio and TV" (Norwegian Sámi Radio website 2006). Rávdná Nils-

Figure 4.6. Rávdná Nilsdatter Buljo, News Director of Sámi Radio, outside the headquarters at Karasjok, Norway. Photograph by Valerie Alia.

datter Buljo, the News Director at Karasjok, told me: "We have our own budget. There are no strings, except we must produce programming in Sámi. There is also some programming in Norwegian. We have daily Sámi radio broadcasts from 7-9 am and 1:30-5:30 pm" (Alia interviews 2006).

By 2003, NRK, Norwegian Sámi Radio, had established an editorial office for indigenous affairs aimed at producing more news spots and features, and programming for the company's nationwide network. In 2006, its website announced the arrival of a news network linked to a range of broadcasting stations of "other indigenous peoples around the world," with plans to expand its website and server "to facilitate exchange of news and topical issues" (Norwegian Sámi radio website 2006). By 2009, NRK Sámi Radio had a MySpace page and an "all Internet" streaming network. NRK Sámi Radio started television broadcasts in 2001. While it is slowly expanding, television remains a relatively small part of output, with only minimal funding and other support from the national broadcaster. Norwegian Sámi Radio produces about twenty hours of programming a year for NRK, mainly about current affairs, children, and youth. Regular programming consists of daily newscasts, and twice-weekly children's programs. There is a daily 15-minute broadcast in Norwegian, and some programs have text translations. "It's still rough," Rávdná Nilsdatter Buljo told me as we sat, relaxing during her coffee break, on a bench in front of the station one late summer afternoon (Alia 2006 interviews).

In Karasjok, it is only a short walk from the Sámediggi—the Sámi Parliament—to Sámi Radio headquarters, and journalists frequent the Parliament and nearby restaurants. The NRK Sámi Radio and Television building was opened in 2000 and still feels new. In a landscaped courtyard outside the building is a sculpted glass version of a traditional Sámi *lavvo* or *lavvu*, a tent-like temporary dwelling. Inside is an assemblage of everyday items, including an iron cooking pot hung over a mock fire. In a prominent position sits an old-fashioned radio. The *lavvo* is sometimes mistakenly called a *tipi* after the dwellings of some Native American and Canadian First Nations peoples. There is a superficial resemblance, with a smoke hole at the top and a framework of poles. A lavvo can be erected by a single person, and is sturdier, to withstand the heavy winds that frequent the Scandinavian tundra. It is widely used as a symbol of Sámi life and culture and forms the basis for the design of both the

Figure 4.7. The glass *lavvo* outside the Sámi Radio headquarters at Karasjok, Norway. Photograph by Valerie Alia.

Figure 4.8. Glass version of a *lavvo* (traditional Sámi tent), together with satellite dish and other up-to-date facilities, mark the Sámi Radio and Television headquarters at Karasjok, Norway. Photograph by Valerie Alia.

theatre and the beautiful Sámi Parliament building at Karasjok. It is not just a relic of the past. While most Sámi have permanent homes, those who herd reindeer continue to use the lavvo as a temporary shelter.

Recalling the struggles of Inuit and First Nations journalists in Nunavut, Northwest Territories, and Yukon to remain independent of Indigenous governments, I asked Buljo whether Sámi Radio operates independently of the Parliament. With a smile, she replied: "Of course we are independent of Sámediggi. We maintain a critical point of view. We can't be the fan club for Sámediggi," she quipped (Alia interviews 2006).

Nils Johan Heatta has served as Director of Sámi Radio since 1985. He was born in Kautokeino in 1954 and worked as a sound and video engineer. As director, he is also a member of NRK company management. He manages a staff of more than 70 and has been active in developing collaborative projects between Sámi radio and television in Finland, Sweden, Norway, and, most recently, Russia. He developed a daily TV news program broadcast in Sweden, Finland, and Norway. In 2006, Heatta told me that although there are "no strings," administrative and financial ties (including television license fees) to the national broadcaster means that NRK limits Sámi broadcast time, which in effect does place strings on the Sámi service. "It is important to have a critical media. Ideally we want to transfer to Sámi controlled broadcast. Ideologically we should be independent. But the question is, is it practical? That we must decide" (Heatta 2006).

On the global front, exchange projects are emerging between YLE, KNR, and Indigenous broadcasters around the globe. "We are working with Maori television in New Zealand which started up in 2005—after only one-and-a-half years they are doing well—and APTN, the Aboriginal Peoples Television Network, in Canada—we are working with them" (Heatta 2006).

In 1999, NRK Sámi Radio started test transmissions on the digital net (DAB), which are available in areas with access to it. While well developed in southern Norway, it is only available in the larger, central communities of Kautokeino and Karasjok in Sápmi. Digital radio broadcasting can also be received by listeners with dish aerials and a DAB radio, and all Sámi Radio transmissions on the digital net can be accessed via NRK Live Radio.

NRK Sámi Radio and SR (Swedish) Sámi Radio collaborate on the joint Internet news site, www.samiradio.org, which publishes current news in Sámi, Norwegian, and Swedish. In 2004, the service added news headlines in English and Spanish. NRK Sámi Radio has its own teletext news pages, with pages set aside for Sámi news and radio and television program schedules. YLE is a co-producer of Nordic Sámi TV News. Sámi Radio in Norway produces television programs for the national broadcaster, NRK. Swedish national television has a Sámi editorial office in Kiruna, which produces about ten hours of current affairs and children's programming for SVT (Sveriges Radio) Swedish television. Program exchange is facilitated by subtitles, digital technology, and the web. NRK Sámi Radio and SvT Sápmi in Kiruna began their daily common news service in 2001, joined by YLE at the start of 2002. Norwegian and Finnish Sámi radio services share an online newspaper, www.samiradio.org, with continuous updates (YLE 2007). Sámi television started with NRK Sámi Radio in Norway, and Swedish SVT (Sveriges Televison) Sápmi, in 2001 and in Finland in 2002. After years of financial challenges, it re-

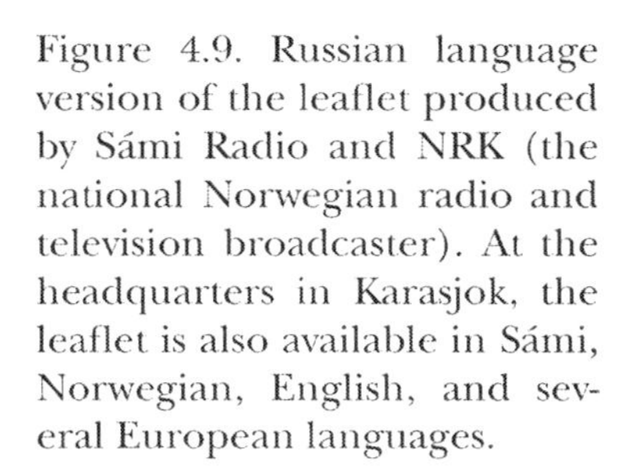
Figure 4.9. Russian language version of the leaflet produced by Sámi Radio and NRK (the national Norwegian radio and television broadcaster). At the headquarters in Karasjok, the leaflet is also available in Sámi, Norwegian, English, and several European languages.

mains under funded, and despite mainstream and international trends, television is still a relatively minor part of the Sámi broadcast day.

Through strategic thinking, careful allocation of human, financial, and technological resources, Sámi have managed to turn some of the disadvantages to advantages, maximizing scarce human and technological resources by sharing them with other stations, in ways that are at once economically expedient, communicatively effective, and culturally appropriate—coming from traditional ways of sharing work and wealth. Sámi Radio services cross the national borders daily via shared transmissions of news, current affairs, and other programming. A digital network and common radio channel are emerging (YLE 2007).

The editorial center of Yleisradio (YLE), Finnish Sámi Radio, is in the village of Inari "where the river Juutua flows into Lake Inari. Programs are made and broadcast in two studios ... The municipality of Inari is the largest ... in Finland." The village of Inari has become the cultural and administrative center for Finnish Sámi, housing the headquarters of YLE, a regional Sámi Educational Centre, the Finnish Sámi Parliament, and Siida, the wonderful Sámi Museum and meeting place. "No other municipality in Finland has as many official languages ... we use North Sami, Inari Sami, Skolt Sami as well as Finnish" (YLE website). Programs are made in three different Sámi languages, "and the journalist of the YLE Channel R3, in turn, makes her programs from Inari in Finnish." Inari is also the nerve center of Nordic program exchange, with programs transferred and exchanged daily with Sámi Radio studios in Sweden and Norway.

Since 1984, Sámi Radio has had its own broadcasting station in Karasjok, Norway, which has twice expanded to accommodate offices, editorial premises, and radio and television production facilities. As a public service broadcaster, the Norwegian Broadcasting Corporation Ltd. (NRK) pledges "to contribute through an active program policy in consolidating and developing Sámi language and culture on a national, regional and county level." NRK plays "a crucial role" in the preservation of Sámi language, culture, and identity. In 2009, Sámi in Sweden will, like those of Norway and Finland, have their own parliament. Perhaps because of the relatively weak support from the Swedish government, Swedish Sámi are leaders in cross-border networking. The "Sámi University College in Guovdageaidnu (Kautokeino) has education in Sámi Media for their students" (Rautio Helander 2006). The more stationary services in broadcast centers are supplemented by local offices at Utsjoki and Karesuvanto, in the municipality of Enontekiö.

In keeping with some of the flexible and inventive practices of Indigenous projects in other parts of the world, Sámi Radio provides year-round access to remote communities through unconventional means: special "broadcasting cars" are fitted with transmitting equipment, turning them into mobile mini-stations. For rougher terrain and weather, especially in mountain regions, the radio transmitters are sometimes carried by snowmobile. Mobile transmitters, carried in various ways throughout Sápmi, allow not just broadcast transmission, but production of on-the-spot live programs. For example, live interviews "convey impressions from the first reindeer round-up of the winter" (YLE 2007). In 1998, Nils Johan Heatta was asked to help Russian Sámi develop a radio station, and two years later, he helped organize Sámi radio broadcasts on the Kola Peninsula. Sámi media are not just expanding services across previously un- or under-served areas of Sápmi, but globally. The photograph following this paragraph is of Juhani Nousuniemi, who has headed Sámi Radio at Inari, Finland, since 1980. Table 4.1 shows the international reach of YLE Radio Finland.

Figure 4.10. Juhani Nousuniemi, one of the visionary leaders, and head of Sámi Radio in Finland since 1980.

Table 4.1. International Transmission of Radio Finland Programming

Country or Region	Language(s)
Northern Europe	Finnish, Swedish, Russian, Sámi
Southwestern Europe	Finnish, Swedish, German, English, French
Southeastern Europe, Middle East and Eastern Africa	Finnish, Swedish
Eastern Europe and Russia	Finnish, Russian, Swedish, English
North America	Finnish, English, Swedish
South America	Finnish, Swedish
Australia and Asia	Finnish, English, Russian, Swedish
Europe	TV Finland Finnish

Source: YLE Radio Finland, 2002.

Figure 4.11. Sámi Radio headquarters at Inari, Finland. As can be seen from the timber construction and the trees reflected in the windows, the structure emphasizes its connection to the land. Photograph by Valerie Alia.

Community radio in India

In 2007, Shri S.K. Arora, Secretary, Ministry of Information and Broadcasting of India, announced that 4000–5000 community radio stations would emerge in "the next few years," thanks to the government's "new enabling community radio policy" and support from UNESCO. He said, "Community radio focuses on low cost and low return … operations, which are aimed at educating and entertaining the community using their own idioms and language in contrast to the private FM radio which is primarily driven by entertainment and business considerations" (UNESCO March 2007).

Mandakini Ki Awaaz (The Voice of Mandakini), the community radio station in Bhanaj, Rudraprayag district, Uttarakhand, India, was founded in 2004 with support from UNESCO and received technical support and training from NGOs. Women play a major role, and in 2007, a women's self help group raised funds for repair and maintenance of the satellite radio receiver originally donated by UNESCO.

An exhibition in New Delhi's Pragati Maidan Exhibition Complex "highlighted the fact that setting up radio stations need neither be complex nor costly" and ran a special live broadcast "Hamari Awaaz FM 89.9 Mhz" from the "Radio-station-in-box." The box measures about 55x50 cm and contains a mixer, CD/Cassette player, 30W FM transmitter, antenna, and a laptop. Using a simple process, the transmitter can be tuned to any frequency within the FM band. Developed by the Asia Pacific Broadcasting Union, with UNESCO assistance the box is a low-cost, "self sufficient broadcasting and production tool." Also on display was a suitcase radio that "can function as a full-fledged radio station." These mini-stations are India's answer to the portable stations used by Sámi Radio and other Indigenous media outlets to bring radio to remote communities.

NITV, Australia's emerging National Indigenous Television Service

On Friday, 13 July 2007, Australia's first 24-hour Indigenous television service was launched in Sydney. The government-funded National Indigenous Television (NITV) broadcasts to about 220,000 people across Northern Australia, Queensland, and South Australia on the Optus Aurora satellite and a channel on Imparja. This development is tempered by years of politicking that some feel will make collaboration among Indigenous broadcasters more difficult. Amidst the celebrations, some people in remote communities expressed concern about their possible exclusion from the new channel. Rita Catoni, general manager of Warlpiri media in Yuendumu Northern Territory, expressed concern that a number of cultural videos might not be considered up to industry standard. "It's going to be a very different type of content on television, which is a shame really, because they're not mutually exclusive, they could have coexisted ... the landscape of Australian television has changed with the launch of NITV. This is an exciting time and a historic milestone in not just broadcasting but Indigenous history" (ABC July 13 2007, unpaginated).

NITV serves about 220,000 people, mainly in remote communities, with plans to extend the service. It aims to "acquire and commission a range of programming which reflects the diversity of Aboriginal and Torres Strait Islander cultures and communities, to support locally produced content, and help to further open up career paths for Indigenous people in the industry." The Chairperson of NITV's board of directors is Professor Larissa Behrendt (of Eualeyai/Kamilaroi descent), Professor of Law and Director of Research at the Jumbunna Indigenous House of Learning, University of Technology, Sydney. She has written on property law, Indigenous rights, dispute resolution and Aboriginal women's issues, and is author of *Achieving Social Justice: Indigenous Rights and Australia's Future*, a Museum of Contemporary Art Board member, and a Director of the Bangarra Dance Theatre.

Until NITV went to air in 2007, Indigenous programming totaled less than two hours a week—just over one percent of total airtime. It began formally in 2004 with the formation of a voluntary committee. A summit in Redfern, Sydney, was "organised by a group of Aboriginal and Torres Strait Islander media professionals and community members committed to the establishment of a national Indigenous broadcasting service." A year later, the Federal Government announced $48.5 million in funding. An NITV media release said, "National Indigenous Television Limited (NITV) has been 25 years in the making and is offering something very different to the average fare on our TV sets." "Black TV is here!" said Professor Larissa Behrendt, Chairperson of National Indigenous Television. "Our first broadcast will be available to more than 600,000 Australians and we will premiere some of the best locally produced documentaries, sport and cultural programs. The broadcast day includes arts, children's programs, culture and music, film, drama, education, language, news, sport, current affairs, comedy, and special events (NITV 2007).

NITV has four-year funding from the Australian Government to purchase existing programs and commission new ones. It remains to be seen how the network will survive after that. Imparja Television carried the launch, so that the inaugural broadcast could be seen in every state.

"NITV will continue to broadcast on the Optus Aurora Satellite which is available to around 220,000 people across northern Australia and parts of Queensland and South Australia." It is hoped that the viewing range will be expanded, following negotiations with pay TV providers and digital platforms "to bring our programs into the home of every Australian" (NITV 13 July 2007, unpaginated).

PACNEWS: Broadcasting in the Pacific Islands

In the Pacific Islands, a milestone arrived in 1987, in the form of PACNEWS, the Pacific news agency. Founded by the Pacific Broadcasting Training and Development Scheme (PACBROAD), a regional radio training project funded by UNESCO and the German foundation, Friedrich Ebert Stiftung (FES), it fostered the production of local and regional news. As of 2001, Helen Molnar and Michael Meadows observed that it was managing to survive without outside funding, mainly through subscriptions from within and outside the region. From its home base with the Pacific Islands Broadcasting Association (PIBA), PACNEWS has reciprocal relationships with fifteen contributing member countries: the Cook Islands, Fiji, the Federated States of Micronesia, Kiribati, the Marshall Islands, Niue, Papua New Guinea, the Republic of Belau, the Solomon Islands, Tonga, Tuvalu, and Vanuatu (Molnar and Meadows 2001: 88).

Television gradually arrived in the 1980s, aided by earlier developments such as PEACESAT—the Pan-Pacific Education and Communication Experiments by Satellite founded in 1971 at the University of Hawai'i to service twenty locations in the Pacific. Molnar and Meadows compare PEACESAT to "experiments in the late 1970s in northern Canada," which included the Inukshuk project, installation of the Anik Satellites, and development of the Inuit Broadcasting Corporation (IBC) and Television Northern Canada (TVNC) (Molnar and Meadows 2001: 99; Alia 1999). PEACESAT used the NASA Applications Technology Satellite (ATS-1) to disseminate public information. Molnar and Meadows note that this was "not a Pacific initiative" and express caution about the motives and methods involved in its operation and the extent of its usefulness to Pacific peoples.

Papua New Guinea started receiving television (mainly from Australia) in 1982 over the INTELSAT 4A satellite, made possible by its repositioning over the Pacific. Two years later, Fiji established the Fiji National Video Centre (FNVC), which produced television programming for children and adults, including a one-hour news magazine, sport, entertainment, and local news. Despite its earlier head start, television did not fully arrive in Fiji until 1986, having been welcomed gradually and cautiously. The Solomon Islands established their own National Video Centre in 1991. This, too, parallels developments in Canada, where the remote Inuit community of Igloolik held out against "southern" TV until the Inuit Broadcasting Corporation was formed—and thus, culturally relevant and Inuit-produced, television was available (Alia 1999; Molnar and Meadows 2001: 103–121).

In the early 1990s, Television New Zealand (TVNZ) developed a small, low-cost television model. It was adopted by four Pacific countries—Niue, the Cook Islands, Nauru, and Samoa, and had a strong Indigenous presence of mainly Māori and Polynesian people (Molnar and Meadows 2001: 122). Tonga has a private television station with limited Indigenous content, run by a Hawaii-based Christian organization, as well as the public Tongan Broadcasting Commission (TBC) station. Vanuatu got its first television service in 1992. Most of Vanuatu's video production originates with the Cultural Centre; Indigenous content is at a fairly early stage of development. In Samoa, a TVNZ model was used as a prototype for developing Televise Samoa, which began broadcasting in 1995. With fewer resources, Kiribati and Tuvalu remain dependent on radio and print media, and on community access video hire (Molnar and Meadows 2001: 140–144).

The Pacific Telecentre Online Community (PacTOC)

In 2006, community telecenter project operators from around the Pacific launched a regional online community website "to support and develop the telecentre movement in the Pacific. The Pacific Telecentre Online Community (PacTOC) gives a voice to grassroots telecentre projects, so they can share experience and expertise with each other and the world." Telecenters "provide a community gathering place where people can access communication technology and applications, learn new skills, tackle local social issues, face common challenges and empower their neighbours" (UNESCO 2006). Seema Chand from Femlink Pacific Media Initatives for Women is editor of the PacTOC website. As she emphasizes, the website enables women's voices to be "heard around the Pacific and the world" and is the "first regional website in the telecentre.org network, a global initiative aimed at supporting community-based telecentres, which emerged from a workshop coordinated by the New Zealand National Commission for UNESCO." The site offers news and events, a resource center, a database of telecenter best practice models and other information relevant to communities in the region, and opportunities for discussion and comment. It was developed with support from the Foundation for Development Cooperation (Australia), the New Zealand National Commission for UNESCO, New Zealand Aid, the Sasakawa Pacific Island Nations Fund (Pacific Island Digital Opportunity Study Committee) Japan, and the telecentre.org program. The parent project, Telecentre.org, is a $21 million Canadian collaborative initiative based at Canada's International Development Research Centre with support from Microsoft Unlimited Potential and The Swiss Agency for Development and Cooperation. It was launched at the United Nations World Summit on the Information Society in Tunisia in October 2005 and includes telecenter networks in India, Sri Lanka, Uganda, Mozambique, South Africa, Chile, and the Americas (UNESCO 28 February 2006, unpaginated).

On 3 May 2007, at the annual celebrations of World Press Freedom Day, the twin themes were press freedom press and journalists' safety. In the Pacific, Indigenous journalists discussed the escalating dangers and hostility toward the media. In one of the main Pacific hot spots, Papua New Guinea, journalists met with government and business representatives at The Divine Word University in Madang. The event "commenced with a peaceful rally through Port Vila town with banners and placards bearing slogans that called for government support for freedom of information and media freedom in Vanuatu." In Fiji and Samoa, the celebrations focused on media freedom. The UNESCO Office for the Pacific States in Apia, Samoa, hosted a program with a keynote address by Savea Sano Malifa, publisher of the award-winning *Samoa Observer* newspaper. At an Editors Forum hosted by The Journalist Association of [Western] Samoa (JAWS), editors called on the Samoan government to enact Freedom of Information legislation, and challenged Samoa's Printers and Publishers Act 1992, which requires editors to reveal their sources (UNESCO 14 May 2007).

A workshop held in May 2007 in Honiara, Solomon Islands, stressed the need to entrench principles of gender equality in media standards. The Workshop for Advancing the Pacific Women in Media Action Plan, organized by UNESCO and the Secretariat of the Pacific Community, sought to bring media practitioners together, to consider how they make decisions on equality and human rights issues, and to discuss media responsibility for raising awareness of human rights violations and promoting human rights education. Participants said that media guidelines, standards, and practices should be consistent with gender equality commitments. They asked that the Pacific Islands News Association (PINA) establish an annual award for the best coverage of gender issues and help its members to develop gender equality policies (UNESCO 5 December 2007, unpaginated).

In Papua New Guinea, the national Indigenous weekly, *Wantok Niuspepa,* is published in the Tok Pisin language. In Tonga, there is *Taimi 'o Tonga* (which Robie describes as "feisty"), published by Kalafi Moala, "who has campaigned for almost two decades for democracy in the kingdom of Tonga," and the "radical newsletter *Ko'e Kele'a* published by pro-democracy Tongan Member of Parliament" Akilisi Pohiva (Robie 2005: 7). The Tongan media follow the pattern seen in Indigenous Canada and elsewhere, in which journalists often become politicians and there is a relatively fluid relationship between political and media institutions and political and media leaders. "Sir Michael Somare, the 'founding father' of Papua New Guinea and current Prime Minister, was once a militant broadcast journalist who turned to politics and led his country to independence in 1975" (Robie 2005: 7).

The story is not always a happy one. "In New Caledonia, Pastor Djoubelly Wea played an activist/journalist role before assassinating rival Kanak leaders Jean-Marie Tjibaou Yeiwene Yeiwene and immediately being gunned down himself in 1989" (Robie 2005: 8). Radio Djiido "has played a crucial role in the independence struggle in New Caledonia while carrying 'revolutionary' news ... Radio Free Bougainville was ... vital in the Bougainville civil war..." In French Polynesia, an investigative journalist, Jean-Pierre Couraud, regarded as being too radical by the establishment, disappeared in 1997 and is believed in some circles to have been murdered (Robie 2005: 8).

Taiwan's Indigenous Television Network

In 2005, Taiwan Television launched the Indigenous Television Network (ITV), a 24-hour channel, dedicated to broadcasting news and educational programming for, and about, Indigenous peoples. It was designed by Taiwan's Council for Indigenous People, with a requirement that 70 percent of staff must be Indigenous. Because many of Taiwan's Indigenous people live in mountainous regions, the Council for Indigenous People installed satellite transmitters in remote villages. The next step will be to improve access for those who cannot afford television sets. Indigenous people make up two percent of Taiwan's population. According to *The Taipei Times*, ITV is the first Indigenous television service in Asia (Weekly Indigenous News 2005).

The Other Side of the Camera: Indigenous Photographers

The photograph "is widely held to be a record, a piece of evidence that something happened." Where cinema creates a sense of continuing activity and life, still photographs freeze historical moments, "capturing and offering up for contemplation a trace of something lost" (Kuhn and McAllister 2006: 1). Even more than film and video cameras, still cameras have tended to stay in colonizers' and visitors' hands. For a long time, Indigenous photographers remained in the shadows and on the margins of Indigenous arts and media, despite the depth, strength, and extent of their work; it is surprising, considering the relative prominence of Indigenous art. Photographers stand at the crossroads between art and journalistic documentation. Walsh (2006: 23) observes that contemporary First Nations artists "are increasingly turning to photography to re-view and re-vision colonial and postcolonial spaces and places."

The Urban Imagery of Jeffrey Thomas and Greg Staats

The images of "two Iroquoian artists, Jeffrey Thomas and Greg Staats ... subvert dominant colonial and postcolonial narratives as well as visual representations of First Nations peoples" (Walsh 2006: 23). Unlike much of First Nations art production in the 1990s, Thomas and Staats generally avoid using such familiar clichés of "Indianness" as feathers, beads, face paint, teepees, horses, or canoes to communicate Indigenous identity. Both Staats and Thomas live in urban settings, Thomas in Winnipeg and Staats in Toronto. Thomas's *Memory Landscape* "signifies postcolonial urban spaces that are experienced through conflict, dispossession and migration, all of which bring about a sense of impermanence. He has created a series of portraits of his son Bear Thomas that documents the boy's growing up as a third-generation urban aboriginal person ... Thomas avoids the use of strategic essentialism" (Spivak 1995; 1990). Thomas explains, "I use my son as a way of putting myself into the landscape. I am saying that there is nothing here that says that there were Indian people here before, or even that had been by there recently. I use Bear as a way to leave a mark on the landscape" (Walsh 2006: 33). In one image, titled *Culture Revolution*, Bear stands in front of a brick wall with large graffiti letters, CULTURE REVOLUTION, wearing a baseball cap with a reproduction of one of Edward Curtis's iconic photographs of a man called Two Moons. In another, *Indian Treaty no. 1*, Bear stands at a Winnipeg roadside in front of a large moving van, sporting a t-shirt with an image of a classic European painting (of an aristocrat) and the inscription, "Founder of the New World" (Walsh 2006: 35, 38–39).

In 1492, Columbus Sailed...and Landed...Where?

The year 1992 marked the five-hundredth anniversary of the arrival of Christopher Columbus in the Americas. Indigenous peoples from Canada and the United States, who protested against celebrating the quincentenary, gathered their actions under the common banner of "500 Years of Resistance." For the first time, two of Canada's main public institutions—the National Art Gallery in Ottawa and the Canadian Museum of Civilization In Hull, Québec mounted major exhibitions featuring, exclusively, works by Indigenous artists. These were *Land, Spirit, Power,* at the National Gallery, and *Indigena,* at the National Museum of Civilization. The two shows were accompanied by major catalogues, lavishly illustrated and extensively written, which were published as books that became widely available and influential (Walsh 2006: 27).

Representing Women

A photographic exhibition titled, *The Marginalization of Indigenous Women,* appeared in Prince Albert, Saskatchewan and Vancouver, British Columbia in 2007. It featured the work of ten First Nations women from Prince Albert, British Columbia, who were part of a project to photograph their daily lives using a technique called "photo voice," a "participatory-action research method" that Caroline Wang developed in the United States in the 1990s. Lana Cook, one of the featured photographers, lives in an urban neighborhood. She says that women living in her neighborhood "don't have a voice where they can address [issues of] homelessness and poverty and street life, prostitution, addictions. [I] want people to see that [people] that are living in this building—we're just like anybody else. We have hopes and dreams" (CBC 2007).

Majority World: A New Global Initiative

In 2007, a "new global initiative" called Majority World was launched. Its objective is to champion Indigenous photographers "from the Developing world and the global South—the majority world!" Founded by kijiji*Vision the DRIK Picture Library Ltd., Bangladesh, and African Media Online, South Africa, its objective is to bring new images to the mass media along with helping Indigenous photographers sell their work. As David Larsen, Director of Africa Media Online, put it, "We are in the business of righting inequalities in the global media market—enabling the majority world to speak with the weight due to voices which communicate on behalf of the vast majority of the planetary population." The project has support from UNESCO's Artists in Development Programme and the Norwegian Ministry of Foreign Affairs. The images are marketed via www.majorityworld.com, a web-based image library featuring the work of Indigenous photographers (*Red Pepper* 2007: 39).

The obvious intention of the editors of the British "red-green coalition" magazine, *Red Pepper*, was to publicize an interesting and important new project that is giving Indigenous photographers work, credit, and an opportunity to present new viewpoints. It is therefore unfortunate that, in using and promoting photographs distributed by Majority World, the editors of *Red Pepper* failed to credit the photographers who created each of the three images that were published alongside the story. Having just discussed the importance of the new service, the magazine renders the photographers, and all but one of the people photographed, anonymous—an example of journalists stopping just short of getting it right (*Red Pepper* 2007: 39).

Power Houses: Linking Politics to Photographic Art

One show by a non-Indigenous outsider is worth mentioning for its portrayal of an Indigenous leader. In 2006, John van der Woude, a student at the Nova Scotia College of Art and Design in Halifax, Nova Scotia, Canada, organized an exhibition of photographs of the homes of Canada's (provincial and territorial) premiers, titled *Power Houses: Canada's First Ministers*. Van der Woude's premise was that "to get a true portrait of someone in power, one should look at their personal belongings." In Canada, the government leaders do not have official residences, and the collection "depicts an eclectic range of housing." The home of Paul Okalik, then Premier of the Inuit-majority Territory of Nunavut in Arctic Canada, is a trailer (caravan), "a normal dwelling in housing-deprived Iqaluit," but a far cry from the grandeur that surrounds many of his colleagues.

Peter Pitseolak: Images of Inuit and Qallunaat Seen through Inuit Eyes

The Canadian Inuit photographer, Peter Pitseolak, widely known as Baffin Island's first indigenous photographer, was also an artist and a writer. His work reflects a lifelong objective of cultural preservation, by documenting the stories and practices, which in his lifetime were rapidly disappearing. He departed sharply from the view expressed by many Inuit and Qallunaat (non-Inuit) that respect for oral cultural traditions means maintaining them in preference to written texts. Like many Inuit in Greenland (which has an extensive body of writing and publications), he stressed the importance of writing and literacy as companions (rather than alternatives) to oral storytelling and recordkeeping. He expressed great regret that Inuit had not developed written texts earlier in their history, to enable them to document and record their own lives and traditions. In the work published as *Peter Pitseolak's Escape from Death,* he wrote, in a satirical, community-and-self-deprecating tone: "We were stupid. We should have thought of writing on sealskins" (Petrone 1988: xii). He saw himself as a social and cultural historian who was recording a rapidly changing way of life. He wrote in the syllabics which missionaries had developed for Inuit of this region (unlike the Roman orthography used in Greenland) and richly illustrated his work with drawings, carvings, and photographs. The 1975 book, *People from Our Side,* documents his early life. Compiled and published (two years after his death) by Dorothy Harley Eber—a Montreal writer specializing in oral history, photography, and documentary reportage—it comprises his own written narrative and an interview with him conducted and taped by Dorothy Eber.

Peter Pitseolak's narrative is one of the few records of early twentieth century Inuit life and the only account which presents an Inuk's point of view—his story, situated firmly in the context of life in his community. Peter Pitseolak spent his childhood in a traditional manner, in nomadic camps. As an adult, he observed and documented the experiences of Inuit who were hosting an increasing array of missionaries and fur traders from southern North America and Europe, Canadian government officials, educators, and other newcomers and visitors from Outside. His writing and photography portrayed the legal and administrative changes brought by national and territorial political and community leaders, and the negative and positive outcomes of the myriad incursions, shifts, and changes. He welcomed the arrival of schools and community education programs, but was distressed by the devastating effects on Inuit, by the alcohol so many visitors brought with them.

Early on, he became fluent in the syllabic writing system, which missionaries had adapted for Inuktitut, the Inuit language, from a system originally created for Cree people. He kept a diary, and began taking photographs when a *Qallunaaq* visitor asked for help, afraid to get close enough to photograph a polar bear. His daughter, Kooyoo Ottochie, has said that her father's interest in photographing his own people, the *Seekooseelakmiut,* may have been sparked by his early encounter with the filmmaker Robert Flaherty. Flaherty visited Baffin Island in 1913 and 1914 and took extensive still photographs. Pitseolak was about twelve at the time and remembered what may well have been a formative experience. Certainly, his own photography and written history provides an insider's alternative to Flaherty's visitor's-eye view of Inuit camp life. After acquiring his own camera in the early 1940s, he began his career of photographing daily life in Inuit camps—a way of life he knew would be forever changed by the increasing movement of Inuit into settlements and the emergence of schools and wage labor. He and his wife, Aggeok, taught themselves to develop film at home, adapting darkroom techniques to the challenging requirements of the Arctic climate. Aggeok was an active partner, who did a great deal of the developing and printing of negatives and was also at home with a camera. She took the famous picture of Peter Pitseolak standing in the snow beside his beloved "122" camera, in about 1946. The designation "122" reflects his habit of naming cameras by the film they took rather than the brand name.

He is perhaps less widely recognized for his wit and humor. The amusing 1940–1942 painting-collage, *The Eskimo Will Talk Like the White Man*, features eleven standing figures of women and men in South Baffin-styled Inuit dress, painted in watercolor with pasted-in faces photographed variously in black and white or color by Peter Pitseolak and others. One man holds Inuit fishing tools, another holds a "white man's saw and shovel." Tucked into (or rising above) their traditional fur hoods are the faces of white people—known and unknown. The man holding the saw and shovel, smiling broadly and looking ready to begin work on some Arctic construction project, has the face of the Hollywood actor, Clark Gable. Two *Qallunaq* faces—a mother and child—peer out of a traditional *amautiq*, the baby-carrying hood of a woman's parka. In the early years of Nunavut, the subject of this work looms increasingly large—as increasingly, *Qallunaat* find themselves learning Inuktitut and Inuit, the language of *Qallunaat* politics and society.

His work is attributed to the Late Historic Period of Inuit art, in the early days when neither artists nor buyers took the work seriously. Though known primarily as a photographer (the medium which captivated him for much of his life), he merits broader recognition. He began working in watercolor twenty years before James and Alma Houston arrived with art supplies in Cape Dorset, having purchased paints and supplies from a Hudson's Bay Company trader in 1939 (Hessel 1998: 26). After he abandoned paint for photography, there was virtually no experimentation with paint in Cape Dorset until the establishment of the Cape Dorset cooperative in the 1970s. Even today, in the midst of increasing excellence in Inuit video and film (including Zacharias Kunuk's award-winning feature film, *Atanarjuat*), there are few Inuit still photographers. About 2,000 of his and Aggeok's negatives dating from the 1940s, 1950s, and 1960s are housed in the McCord Museum's Notman Archives at McGill University, Montreal. On 15 July 1923, he married Annie in Lake Harbour. In 1941, after her death, he began living with Aggeok, who became his second wife and lifelong collaborator, developing and printing most of his photographs and taking many of her own (though claiming Peter Pitseolak always set the camera). Her brother Johnniebo and his wife Kenojuak were among the first Cape Dorset (Kingnait) artists to achieve distinction.

Images and People in Motion: The Emergence of Indigenous Film

Indigenous film was long dominated by outsiders' representations. Robert Flaherty's *Nanook of the North*, shot on Hudson Bay's Ungava Peninsula in northern Québec and released in 1922, is widely considered to be the first full-length documentary and the first feature-length filmed representation of Indigenous people. It was followed by feature films in which Indigenous people were portrayed by actors from myriad cultural groups—for example, the 1961 Hollywood "Eskimo" film, "The Savage Innocents," starring Anthony Quinn and Yoko Tani.

At the end of the 1960s, Canadian Indigenous filmmakers began to tell their own stories. The first national Indigenous television documentary, *Sharing a Dream*, directed by Michael Mitchell (Mohawk), appeared in 1988. In 2000, *Atanarjuat* (*The Fast Runner*), directed by Zacharias Kunuk—the first full-length feature film produced, written, directed, and acted by Inuit in Inuktitut—won the Caméra d'Or prize for best first feature at the 2001 Cannes Festival. Indigenous filmmakers have created an international network, the First Nations Film and Video World Alliance, whose membership currently includes practitioners from Canada, Vanuatu, Mexico, the United States, Greenland, Australia, Aotearoa/New Zealand, and the Solomon Islands.

In Canada, Michael Posluns praises the "pioneering work" of the "Indian Film Crew," a joint project of the National Film Board (NFB), which provided training and equipment, and the Company of Young Canadians (CYC), which provided subsistence for crewmembers. As I have noted elsewhere in relation to Indigenous leaders and journalism, Posluns notes the number of political leaders who "got their earliest opportunities to speak out through this [film] program." He also notes that their films are no longer listed in the NFB catalogue and wonders whether new technologies and dissemination techniques might help to challenge "the capacity of a corporation to disappear works of historic, political and artistic importance" by making them available in other ways (Posluns 2007).

Carol Geddes, a filmmaker from the Teslin (Tlingit) First Nation in Yukon, brought still and moving images and image-makers together in her 1997 film, *Picturing a People: George Johnston, Tlingit Photographer*, a portrait of a photographer from her own community. George Johnston was self-taught, and worked from 1920–1945, bringing images that linked tradition with contemporary life. Teslin is the only inland Tlingit community, its members having migrated from the west coast of Canada to Yukon, to trade with other First Nations. Johnston first made audio recordings and films. At age 16, he traveled from Teslin to Tlingit communities on the Alaskan coast, recording traditional songs and dances. Geddes' most recent project is a six-part live-action animation series on Tlingit people's interaction with Russians in the 1800s. Geddes' 2004 animation *Two Winters: Tales from Above the Earth* won numerous awards.

Founded in 1999, Arnait Video Productions Collective, based in Igloolik, Nunavut (Canada) calls itself "the first women's collective independent production company in the Arctic." In 2008, it premiered its first full-length feature film, *Before Tomorrow*. Like the films of its coproducers, Igloolik Isuma Productions Inc. and Kunuk Cohn Productions (*Atanarjuat* and *The Journals of Knud Rasmussen*), *Before Tomorrow* was made in Inuktitut, with subtitles available in other languages. Codirected by Madeline Piujuq Ivalu and Marie-Hélène Cousineau, it won best Canadian first feature at the 2008 Toronto International Film Festival. Among Arnait's other films are *Ninguira* (My Grandmother, 1999), about women and health; *Anaana* (Mother, 2002); and *Unakuluk* (Dear Little One, 2006) about Inuit adoption (Arnait Video Productions Collective 2008, online, unpaginated).

ImageNative

The annual ImageNative Film and Media Arts Festival celebrated its tenth anniversary in Toronto, in October 2009. Since its inception, the festival has presented distinguished and often ground-breaking Indigenous films from around the world. The 2006 festival had a particularly impressive lineup. It featured the international premiere of the courageous film, *Tuli*, by Kanakan Balintagos of the Palaw'an nation of the Philippines, about a young girl forced into an arranged marriage by an abusive father, who falls in love with her childhood (female) friend. The closing gala film was *Waban-Aki: People from Where the Sun Rises*, a feature-length documentary written and directed by the distinguished Abenaki filmmaker, Alanis Obomsawin, which documents her return to the village where she was raised. Also screened was Inuit director Zacharias Kunuk's feature, *The Journals of Knud Rasmussen*. The second feature film by the maker of the award-winning *Atanarjuat* (*The Fast Runner*), the *Journals of Knud Rasmussen* made it into the mainstream as well, chosen to open the 2006 Toronto film festival. While Kunuk's first film dealt solely with the Inuit community, *The Journals of Knud Rasmussen* "looks at Inuit experiences of colonial incursions." It focuses on the great shaman, Awa, his daughter, Apak, and the intercultural encounters with Christian missionaries and traders in the early 1920s, using material from Rasmussen's journals and interviews with Inuit elders. Though considered a Danish explorer, Rasmussen was part Greenlandic and spoke the language. It is interesting to note that, although the Toronto festival claimed to present the world premiere, Kunuk is committed to showing all of his work to Inuit first, and the film was first screened, in Inuktitut, to Inuit communities in Canada and Greenland.

He has consistently used his successes to bring more media work to Indigenous communities, and in November 2006, launched the Indigenous Film Network in the Nunavut community of Rankin Inlet, an offshoot of Kunuk's company, Isuma Distribution International, which aims to bring more films to remote Inuit, Métis, and First Nations communities across Canada. "Kunuk says he makes films for aboriginal people but they rarely get a chance to see them." "We're trying to reach small communities that don't have theatres" (CBC 28 November 2006).

Each year's ImageNative festival features work from a different part of the world. In 2006, it focused on the Pacific Islands, with a program guest-curated by the feminist Māori filmmaker, activist, actor, and teacher, Merata Mita. New Media and Radio works have a high profile and their own venue at the festival. Here are some examples of the 2006 offerings:

First Vision http://firstvisionart.com—a website designed by Archer Pechawis celebrating three artists. Archer Pechawis is a "media-integrated performing artist, new media artist, writer, curator, and teacher" from Canada. His work "investigates the intersection of Plains Cree culture and digital technology" (ImageNative 2006).

Inclusive http://interactiveclick.ca—website designed by visual artist Jason Baergfrom Canada. "This project investigates the contemporary urban Indigenous identity. It aims to encourage inclusiveness, community and growth, as well as to offer blog and podcast opportunities."

Wepinasowina www.wepinasowina.net—website by Cheryl L'Hirondelle (Canada), "part of Cheryl's ongoing commentary about identity. As an Apihtawi-Kosisan Iskwew (halfbreed woman) with a Nehiyawin (Cree worldview) gaze, her research and experience is that there is not any single flag, design or emblem that denotes a sense of collectivity or nationhood." Cheryl L'Hirondelle (a.k.a. Waynohtew, Cheryl Koprek) describes herself as "a Vancouver-based halfbreed (Métis/Cree-non status/treaty, French, German, Polish) multi/interdisciplinary artist"(ImageNative 2006).

Kiss My Black Arts is a live Indigenous arts program that airs weekly on Koori Radio 93.7. ImageNative featured a segment by Gaven Ivey (a.k.a. Naian), an "Aboriginal South Sea Islander from the far north coast of New South Wales, Australia." Its subject is the Indigenous Gay and Lesbian events of Mardi Gras week in Australia, which included the Boomalli exhibition *Pink, Black and Beautiful* (ImageNative 2006).

Native Vibes is a US one-hour weekly radio series hosted by Charles Shoots the Enemy (Hunkpapa Lakota, South Dakota, USA) featuring contemporary Indigenous music and commentary (ImageNative 2006).

Ute Mountain Tribal Park—Mesa Verde Centennial Birthday, a "guided audio tour through the Ute Mountain Tribal Park by Kim Pappin, a member of the Osage Nation from Pawhuska, Oklahoma who was raised in the US Pacific Northwest. She now lives in Colorado, where she is Music Director and Producer for KSUT 91.3 FM—Southern Ute Tribal Radio" (ImageNative 2006).

Also in 2006 came Australia's first feature film in an Aboriginal language, Rolf de Heer's *Ten Canoes*, which won six awards at the Australian Film Industry awards in Melbourne (CBC 8 December 2006).

Sámi Film

In 1996, Sámi launched an annual Sámi Film Festival in Guovdageaidnu (Kautokeino), Norway. The 2008 festival opened with the premiere of Nils Gaup's feature, *The Kautokeino Rebellion*, which revisits historical events from a Sámi perspective. Gaup is perhaps the best known Sámi filmmaker, and the first to achieve prominence on the international scene. His 1988 feature, *The Pathfinder*, was nominated for an Academy Award.

The Aboriginal History Media Arts Lab

Not all progress is future-directed. Some of the developments involve reworking, or re-presenting history. On 22 February 2006, one of Canada's pre-eminent Indigenous filmmakers, Loretta Sarah Todd, launched the Aboriginal History Media Arts Lab (AHMAL) with a screening of the 1930 silent film, *The Silent Enemy*, a fictional account of traditional life in northern Canada, with an all-Aboriginal cast. Using "an old film to launch the lab allows us to speak back to a cultural artifact from Hollywood, intervening with contemporary Native music that both comments on the stereotypes and responds to the dramatic elements of the story." The accompaniment she refers to is a live performance by Russell Wallace "and friends" of a score fusing "jazz, classical, electronic and traditional native music." AHMAL is an independent organization that was founded to promote "Aboriginal new media and technology-based art and communications to a wide audience." It has "formed strategic partnerships" with the Chief Dan George Centre, Simon Fraser University, and the First Nations Studies Department at University of British Columbia, chosen for "their leadership in media issues and technologies." AHMAL looks simultaneously backward and forward, seeking to correct the historical record and "counter persistent misunderstandings about Aboriginal societies," while producing innovative media to "improve the quality and quantity of Aboriginal media arts." The Lab is "a forum to promote the complexity of Aboriginal people by promoting and encouraging Aboriginal media arts that renew, reframe and assert our histories—and our presents and futures" (Todd 2006). Todd said the "long-term vision is to become an actual Media Lab offering a place for experimentation and the practical application of technology in the lives of Aboriginal people and communities" (Todd 2006).

In the next chapter, we will look at the current trends and future directions of Indigenous media networks and projects.

CHAPTER 5

We Have Seen the Future

"Standing with Legs in Both Cultures"

Were he still alive, my father-in-law, the great journalist and "muckraker," Lincoln Steffens, might express both delight and horror at the array of puns and paraphrases, quotes and misquotes, permutations and misrepresentations that have arisen from his original utterance, "I have seen the future and it works," which he also expressed as, "I have gone over into the future." I hope he would forgive me for yet another appropriation of the phrase. This concluding chapter looks at the contemporary realities and future potential of the New Media Nation. With tongue firmly in cheek but a serious point in mind, Lincoln Steffens declared philosophy a waste of time:

> No more philosophy for me. There was no ethics in it.... I had been reading [philosophers who] thought they had it all settled. They did not have anything settled ... they could not agree upon what was knowledge, [or] what was good and what evil, nor why (L. Steffens 1931: 139).

The ethicist, Clifford Christians, offers some clues to good, evil, and the potential usefulness of philosophy. He wants an "ethics of justice" with media distribution based on need. He says that "global mass media are not neutral purveyors of information, but creators and shapers of culture ... institutional agents of acculturation..." and the "electronic superhighway cannot be envisioned except as a social necessity" (Christians 1989: 3). I think Steffens would have appreciated that. I think he would also appreciate the work of the Nishnawbe (Ojibway) spiritual teacher and poet, Arthur Solomon, whose poem, "War Song," links everything to action:

> *There is no middle ground,*
> There are many people who have seen the way things are,

> And have asked almost in despair,
> But what can I do?
>
> And the only answer has been,
> *You* have to do something about *You*
> Only you can decide whether you will be a part of
> This destruction or whether you will set your
> Heart and mind against it.
>
> You may not be able to change where you work or how
> You earn your living.
> But you are totally responsible for the direction that
> You give your own life.
> ... whatever we are, we must be action people (Solomon 1990: 67).

Art Solomon challenged us to commit ourselves to transforming those philosophical meanderings that so frustrated Lincoln Steffens into action, much as Steffens himself turned to using journalism for social change. Today, new forms of muckraking and transformative journalism are emerging. I think if Steffens were alive, he would join the ranks of enthusiastic bloggers, and would support and celebrate Indigenous media works and networks. Indigenous people are challenging predominant practices and values. The boundaries they are pushing, and crossing, are not just geographic.

> The boundaries of poetry, plays, song writing, fiction and non-fiction are blurred as indigenous writers seek to use language in ways which capture the messages, nuances and flavor of indigenous lives. The activity of writing has produced the related activity of publishing. Maori newspapers, which were quite common in the nineteenth century, have been revived as different organizations and tribes seek to provide better information than is available in the mainstream media. Language revitalization initiatives have created a demand for multi-media language resources for children (Tuhiwai Smith 1999: 150).

As we have seen, Indigenous media are increasingly bringing such resources to wider audiences.

Two Steps Forward, One Back...

It is a mistake to assume that self-determination, self-government, and Indigenous control of media are inherently progressive. It is unfortunate that communities that are already marginalized sometimes promote the marginalization of others.

In 2007 in the United States, the Cherokee Nation voted to revoke the tribal citizenship of about 2,800 descendants of former slaves "owned" by Cherokee people. Ironically, the decision relies on tribal membership rolls imposed by the US government as a divide-and-conquer tactic in its effort to divide and reassign Native American lands. Nearly 80 percent voted in favor of an amendment to the tribal constitution restricting citizenship to descendants of "by blood" tribe members as listed on the century-old official government rolls. Some members of the Cherokee Nation called it racist and an example of the worst form of identity politics. Marilyn Vann, president of the Descendants of Freedmen of Five Civilized Tribes in Oklahoma City, said: "I'm very disappointed that people bought into a lot of rhetoric and falsehoods by tribal leaders." She called it a "fraudulent election" (CBC 4 March 2007 [online, unpaginated]).

In June 2006, Pete Steffens and I traveled to northwest Washington State (Lummi Nation 2006). Pete spent many years teaching journalism to people from the Lummi and other Native American nations, organizing projects and programs among Indigenous journalists, and exchange programs with Indigenous and non-Indigenous journalists, journalism students, and teaching staff at Northwest Indian College and Western Washington University. The Lummi, a Coast Salish people in Washington state, are known as politically astute and active, having successfully lobbied for Indigenous rights in the US and internationally and developed programs ranging from a fishery, a university, and a casino. At the Stommish, we met old friends and colleagues, including Lummi Nation President, Darrell Hillaire; Fred Lane, a journalist, artist, editor of the Lummi newspaper, *Squol Quol* and producer of broadcast programming and community events; and Jim Broder, who works with the (Salish) Tulalip tribe (which, as we noted earlier, was the first tribe with its own cable television station).

When we lived nearby, in Bellingham, in the 1990s, we watched the newspaper develop, and were participant observers and participants in the process of facilitating Indigenous media networks in the region. In those days, the focus was on newspapers and community newsletters. A decade later, we were given a DVD containing an edition of *Northwest Indian News*, referred to several new websites and multimedia experiments, and told about the advent of "Native American TV." "We're trying to get on the air," Fred Lane told us, adding that the television service was just starting. It covers Washington state and other parts of the northwest coast of the United States, carried by the Seattle station,

Figure 5.1. Fred Lane (on right), Lummi Nation editor and multimedia journalist, with Pete Steffens at the 2006 Stommish water festival, Washington, USA. Photograph by Valerie Alia.

KVOS. In 2008, Pete and I were watching APTN at home on Vancouver Island, British Columbia, just over the border from Washington State, when an edition of *Northwest Indian News* came on the air. Within two years, the project was not only reaching Native American communities across the northwestern United States, but had crossed the border into Canada to become a feature on APTN.

In this region, there is much borrowing and sharing across cultures, including in the Stommish itself, which began as a Salish water festival. Amidst the war canoe races and other events derived from Lummi traditions, a pan-Indigenous mini pow wow has emerged in recent years, with dancers, dance styles, and costumes of central and eastern plains and woodland peoples, and a scattering of stalls with crafts, CDs, and other items from Lummi and other Indigenous sources. In the midst of the multigenerational, multicultural fun, there were occasional lapses. Along with this attitude of joyful and respectful cross-cultural sharing are expressions of less tolerant and sometimes unfriendly feelings. I overheard a conversation between a young Lummi woman and man who looked as if they were enjoying the festivities. As they passed one of the concession stands, which was run by Peruvian Indigenous people selling crafts, the woman said loudly, in a derogatory tone, "I won't buy from *them*. I'll buy things from my *own* people." Her tone implied that "they" were imposters, though she did not protest the non-Lummi pow wow dancing that was taking place nearby. In truth, there were very few Lummi crafts for sale, and all of the Stommish vendors have to pay fees to the Lummi Nation and gain official approval to exhibit there. And, like most of the others, these vendors were also Indigenous people.

The Stommish experience highlights a continuing tension between Indigenous solidarity/pan-indigeneity and cultural chauvinism. Neo-conservatism exists alongside the more progressive aspects of cultural revival. The young woman's outburst had overtones of racism, elitism, and classism. Her position of "Lummi first and only" ignored the cultural appropriation of other Indigenous cultural practices taking place in front of her. The incident underscored the fact that not all intolerance and racist sentiment comes from white colonials. Nor do all First Peoples follow egalitarian principles. Only a few generations ago, the Lummi owned slaves, and today, people still make comments (often tongue in cheek) about whose ancestors were aristocrats and whose were slaves. None of this negates the horrors of colonization or the historic and contemporary oppression of Indigenous peoples.

But double standards carry dangers of their own. Oppression and denigration of one Indigenous nation or other group by another cannot be seen in a lesser light. I remembered this incident when a couple of First Nations people I know and respect said that, despite acknowledging his vicious anti-Semitic and other remarks, a prominent First Nations man should be dealt with informally by "his own people" and not held accountable in the public arena, as one First Nations colleague told me in a confidential email. In the United States, the Seminole and Cherokee tribes have disenfranchised former tribal members with black/African American ancestry. Ironically, in citing "authenticity" of tribal membership and tradition, they rely on membership rolls that were originally imposed on them by the US government and its Bureau of Indian Affairs (Staples, B. 2003: 9). Like the Lummis, Seminole, Creek, and Cherokee peoples kept slaves, some of whom were part African American.

The Brilliance and Bravery of Everett Soop

Sometimes, the messenger is killed or muzzled at home as well as abroad. The brilliant journalist, public speaker, and editorial cartoonist, Everett Soop, called himself "the pit bull terrier of native journalism." Soop was an equal opportunity muckraker. He told the journalist Sandy Greer, whose 1998 documentary, *Soop on Wheels*, celebrated his life and work, that "self-determination requires healing, and healing means no longer pushing unpleasant realities under the carpet." For his dedication to truth-telling, he was "ostracized by many people in his own community" (Greer 1999: 39). In one of his best-known cartoons, a First Nations man sits astride a horse, in full Hollywood-style regalia complete with buckskins and feathered headdress. He has turned in the saddle to look behind him, where he (and we) can see that several arrows have pierced his back, and one has landed on the posterior of his horse. The caption reads, "I see my tribe is still behind me" (Greer 1999: 39). Giving equal time to the troubles beyond his "tribe," another of Soop's cartoons depicts a disembodied hand tearing a piece of toilet paper from a roll attached to an invisible wall. On the main part of the toilet paper roll is the text, "NEW Constitution." The piece being torn off reads, "INDIAN TREATIES" (Greer 1999: 41).

Soop was one of Canada's first Aboriginal columnists and cartoonists, joining the newly launched *Kainai News* (one of the first Indigenous newspapers in Canada) as an editorial cartoonist and columnist in 1968. His work has appeared in numerous other Aboriginal and "mainstream" publications, and he is the only Indigenous artist to be included in the Museum of Caricatures of the National Archives of Canada. He said, "I am not an expert on Indians, I am an Indian, and I have a right to speak my truth as I live and experience it" (Greer 2001). That truth was enriched by a wide range of commitments, passions, and interests. In addition to his work as a journalist and an artist, Soop was a political leader during the 1960s. His art and politics were infused with pride in his Blood Indian heritage and membership in the Blackfoot First Nation. He spoke Blackfoot fluently and "paid close attention to traditional teachings and practices," and also loved classical music. The "studio in his home on the Blood Indian Reserve in Southern Alberta was filled with books on philosophy, anthropology, art and world religions" (Greer 2001).

When he was diagnosed with muscular dystrophy at 16, he thought his life would soon end. Instead, he lived to age 58, spending most of the years energetically challenging people and policies, both within and outside First Nations communities. "Outrageousness was his way to challenge adversity ... he believed his important work in life did not begin until he was in a wheelchair—until he had become an advocate for the disabled." He said, "I never did really see myself as a cartoonist. I saw myself as a humorist. So the way I viewed muscular dystrophy was almost incidental to what I wanted to do. I guess it stopped me from doing a lot of things ... To me, life is all about people and the world around you, not what you become." When he lost the use of his legs about 20 years ago, he faced what he feared most—becoming what he called a "vegetable Soop" in a wheelchair. Instead, he worked diligently to educate medical professionals, the wider public and peers with disabilities (Greer 2001).

In words that echo those of Conway Jocks, Everett Soop declared, "A cartoonist is not a crusader, but rather a tiller of the field where the battle is fought." Indigenous media practitioners are fighting on many fronts. As we will see below, the Māori videographer, Dean Te Kupu Hapeta, and others are dedicated to what some of them are calling nonviolent media "warfare."

Waging War Nonviolently: Indigenous Media Projects

In the 1960s, Gil Scott Heron sang, "The Revolution Will Not be Televised." Cristina Verán writes that Indigenous peoples are making "community-based and produced video news media—a revolution all its own." "Some initiatives are tribe-based, others international collaborations, still more are the brainchildren of individuals breaking down barriers and documenting the process all the while, behind and in front of their own cameras. Indigenous issues, struggles, and triumphs *will* be televised" (Véran 2006).

Māori videographer, Dean Te Kupu Hapeta, says, "Media is our nonviolent way to wage war." He is the producer of *Ngatahi: Know the Links* (*ngatahi* means "united together" in Māori), a video documentary series featuring Indigenous and other youth. Twenty countries have been included. The project is supported by the videographer's *iwi* (tribal groups)—Ngati Raukawa, Ngati Toa, and Te Ati Awa, and Creative New Zealand. Hapeta is a pioneering hip-hop artist as well as a videographer, "fronting the politically charged Upper Hutt Posse." The videos merge interviews with music in "a kind of 'rapumentary,' embodying the language and aesthetic favored today not only by Māori youth, but Native young people in many parts of the world. What is crucial is that the

Figure 5.2. The Māori filmmaker, Dean Te Kupu Hapeta, on location in Gugulethu Township near Cape Town, South Africa, filming a documentary for his series, *Ngatahi: Know the Links.* His subjects are the South African hip-hop artists, The Kronik Crew. Photo by Mustafa Maluka for the Fellowship of Reconciliation.

series 'comes from an indigenous perspective, presenting other indigenous and marginalized peoples' perspectives."

These experiments in hip-hop are part of a worldwide explosion of what Abraham Bojórquez, leader of the Bolivian group, *Ukamau y Ke,* calls "'a revolutionary genre.' In the wake of the election of Bolivia's first Indigenous president, Evo Morales (and the earlier election of the Indigenous President, Hugo Chávez, in Venezuela), young musicians and hip-hop artists have an ever-expanding audience for their blend of "ancient Andean folk styles with politically charged rap lyrics espousing 'liberation.' Bojórquez says, 'people listen to us.'" (Carroll and Schipani 2009: 34). There is a similar mix of traditional and innovative style and content in the work of the Canadian duo, *Urban Spirits Dance.* Their 2009 visit to Nanaimo, British Columbia, was typical of their community-based way of working. Holly Bright, the visionary artistic director of Nanaimo's *Crimson Coast Dance Society,* invited a group of young people to help organize Body Talk, a free annual event that features guest artist-teachers and encourages community-wide participation. *Urban*

Spirits Dance consists of James Jones, "aka Bboy Caution," and Angela Miracle Gladue, "aka Bgirl Lunacee." The two have performed and taught internationally, drawing on a repertoire of hip-hop, breakdancing, funk, and traditional Métis and First Nations dance. "The intergenerational nature of the workshop serves to build bridges between age groups, neighbourhoods and the community as a whole" (Nanaimo Bulletin 2009).

In the 1980s, the Inuit broadcaster and political leader Rosemarie Kuptana famously compared non-Indigenous television to the neutron bomb, which destroyed the soul of a people while leaving "the shell of a people walking around. This is television in which the traditions, the skills, the culture, the language, count for nothing" (Kuptana 1982: 6). She made this comparison at a time when Indigenous people were campaigning widely and strongly for the resources, training, and facilities with which to develop and run their own media. The statement was taken up by Indigenous media organizations around the world, as an expression of the relationship between media and cultural genocide. I have heard it quoted in Greenland, Australia, and New Zealand, in the northern and southern United States, and in the Sámi countries. At the time, Inuit were experiencing and protesting the effects of "mainstream" television on people in remote Arctic communities, many of whom still relied on a subsistence economy. They were suddenly inundated with (among other things) programs and advertisements telling them to buy expensive cars, elaborate clothes, and electrical appliances. Thanks to Indigenous programming and networks, today's audiences also receive images of cultural survival and revival. The following are some examples of current and forward-looking projects.

In Tanzania, a group of young people in the city of Arusha founded *Aang Serian*, ("House of Peace" in the Ki-Masai language), to promote Indigenous arts, culture, language, and traditional knowledge. In 2002, the group developed *Aang Serian* Drum, a community-based Indigenous media project that provides local video training, education, technological assistance, and equipment to individuals and villages.

People (and peoples) in remote areas often leap frog from very old technology to the new technology while those in the "developed world" muddle along. The rapid spread of cell phones in China and other rapidly developing countries is an example. We carry on with wired phone systems because we have the technology on hand and the rate for cell phones is much higher, at least in areas with unlimited calling on land lines (Posluns 2007):

> As Internet accessibility increases and the costs of DVD reproduction decreases... the route from camera to editing room to audience, peoples to peoples, opens up still more avenues. Partners of the international Indigenous Media Network are gathering for an annual convocation at the United Nations Permanent Forum on Indigenous Issues ... to share strategies for increasing their voice and visibility on the world stage. Armed with increasingly affordable and accessible recording technology, fiscal support, and online, broadcast, and other distribution mechanisms ... Indigenous media arts, actions, and activism have an increasing ability to speak to and influence their world—and yours (Verán 2006).

> The pan-indigenous element is essential in reviving pride and in fostering a sense of commonality, both of which can go on quite nicely side-by-side with cultural revival. If there were a solid, well-rounded curriculum in First Nations schools these themes would alternate. ...George's stories about Andrew Paull organizing lacrosse tournaments across the country in order to bring communities together at a time when political activity had been criminalized by Parliament. Lacrosse was, of course, a culturally specific custom...bringing leaders, parents and young people together and providing them with a means of communication and a sense of pride laid a foundation on which those same people could carry on to re-build their own cultures. One of the great obstacles most non-Indigenous people have in understanding Aboriginal and treaty rights ... is that they are collective rights belonging to communities. I think that this has not only great legal significance, but that it is also the crux of cultural understanding. Even today, the Government insists on land claims being settled within a given province even though, for example, the James Bay Cree did not historically notice any boundary between their lands in what is presently Ontario and what is presently Québec (Posluns 2007).

In 2007, the United Nations Permanent Forum on Indigenous Issues met in New York and launched an Indigenous Commission for Development of Communications Technologies (ICTs) in the Americas. Based in Ottawa, Canada, it is headed by a board of directors comprising representatives from all of the participating countries. Its president is Tony Belcourt, President of the Métis Nation of Ontario (Canada); other officers include Vice-president Jayariyu Farias Montiel, Director of Periodico Wayuunaiki of Venezuela, and Pedro Victoriano Cruz, Director of Xiranhua Comunicaciones of Mexico (ICCTA 2007).

Following the exhibition and government announcements in India, described in Chapter 4, an experimental project titled Telecentre on Wheels was launched. The Telecentre brings information on health and hygiene, literacy, adult education, agriculture, human rights, and civil laws in the Bengali language, to people in remote villages in West Bengal, India. The project was organized by Change Initiatives, a rural Indian NGO, with support from UNESCO and the West Bengal Renewable Energy Development Agency (WBREDA). Change Initiatives had spent five years working with rural communities in the field of ICT for development in the Indian state of West Bengal. The project began with the Nabanna Information Network, with support from UNESCO's International Programme for the Development of Communication (IPDC). Its objective was to develop knowledge networks to facilitate information sharing by rural women. It is no surprise that the research showed that improving ICT access "could speed up development." But such projects need to take into account the lack of electricity and connectivity in India's remote areas, which inhibit or prevent ICT access. With this in mind, the first step was to introduce laptops in two villages, Ghoragacha and Madandanga. The next step was to set up a mobile information kiosk that travels among the different villages. Still, the problem of electricity and connectivity persisted. In mid 2007, Change Initiatives began working with WBREDA, which had developed a solar-rickshaw van for bringing diesel fuel to villages. The solar van model was adapted to mobile telecentres, which could bring ICT tools to villages without the need for continuous availability of power. This became the Telecentre on Wheels (TOW). "TOW is a customised tricycle (rickshaw) equipped with a solar panel and necessary hardware, such as laptop computer, printer, power panels, etc. After customisation and testing, TOW was formally launched in November 2007. TOW now covers four villages, Ghoragacha, Madandanga, Kantabelia and Teligacha, and travels from one village to another on a regular basis" (UNESCO 18 December 2007). The program includes training a resource person and several people from each village.

TOW aims to develop a financially sustainable model by providing various ICT-enabled services and by helping local women to sell their products. The initiative also builds up a digital archive of local content on the issues concerning women, youth, and rural livelihoods, which allows villagers to access relevant information at their doorstep. Using Ethnographic Action Research (EAR) techniques, this project determines local needs and examines the transformation in village life through ICT.

In 2006, Iqaluit (once known as Frobisher Bay)—the capital of Nunavut—took another step into an already cyber-savvy future. Previously, the community of seven thousand was well served by a single hotspot—a free wireless Internet access point—that covered most of Iqaluit. It was provided by a two-person computer networking shop, with a hotspot on top of the eight-story building that is the town's tallest. ComGuard CTS offered the service free "to anyone within sight of the antenna" and will expand "according to demand." Iqaluit residents who can afford the fee also have high-speed Internet access through either Northwestel, the telephone provider, or the Qiniq satellite network run by the Nunavut Broadband Development Corporation (Minogue 2006). As Minogue reports for the CBC, the wi-fi network is "the latest leap in a city that is increasingly wired." Increasingly, politicians and others are seen with personal digital assistance and cell phones (mobile phones). With a typical mix of tradition and innovation, the reporter observed: "On Tuesday, a man interrupted a harpoon-throwing contest, part of an annual week-long spring festival, to answer a [mobile] call" (Minogue 2006).

In order to use the free service, people need to have wireless network equipment installed on their computers, and "can access the World Wide Web through ComGuard's servers, connect through a personal digital assistant or other portable device." But scarcity of private businesses and businesspeople requires alternative service provision. ComGuard enlisted the help of the Nunavut government and is installing more antennae in cafes "to lure web-surfing customers" (Minogue 2006). As in other Indigenous communities around the world, access is a creative mix of public and private, with the sharing of resources a dominant theme. As in other remote locations, wiring Iqaluit means addressing special conditions. Like other northern towns, Iqaluit has more ravens than people. It was necessary to install a new transparent dome over part of the hardware on the high-rise building, after a group of the twenty-pound birds had eaten away the original cover (Minogue).

And in 2005, in another part of the world, Nibutani, Japan hosted the Global Congress of Indigenous peoples. Nibutani is the home of Ainu leader Shigeru Kayano and the Ainu FM radio station, *Pipaushi* ("place rich in shells"), which he founded. Its director is his son Shiro. *Pipaushi* broadcasts globally on the Internet, often playing the music of the Ainu activist and songwriter, Kanô Oki (Temman 2006 [unpaginated]).

"There was a time when the Ainu culture seemed nearly extinct, drowned in the long-standing Japanese assimilation policy. Now a new wave is swelling. A tiny community FM radio station [*Pipaushi*] that opened five years ago [2001] in Biratori, southern Hokkaido, is keeping the Ainu voice alive and reaching audiences on a global scale." According to Eiji Zakoda, the community of Nibutani "has become a global gathering place for the international movement to protect indigenous culture."

"Listen to our radio broadcast and you can actually hear the language that has been revived," Kayano said in an interview shortly before his death. "Words fly to you through the air. It is my dear wish that the Ainu words will reach listeners and tell our story." He got the idea for the station after learning that Indigenous people in Canada "have radio programs in their own language." He located the station at Nibutani library; it went to air in April 2001, with the support and involvement of Kayano's son, Shiro. Its official name is FM-Nibutani but "people call it FM-*Pipaushi*" (Zakoda 2006). While the community FM station is available only locally, as in the case of Native American radio and many others, the Internet has extended its access, worldwide. In 2006, Shiro's International Indigenous Peoples' Network Website was receiving about 300 visitors for each broadcast. "It is a great idea to promote the Ainu language through the Internet," commented a Japanese resident of Seattle, and a high school student in Germany wrote: "I want to learn about the Ainu." Zakoda said that these and other comments were reflected on one of the station walls, which has been "covered with messages from various indigenous people around the world" (Zakoda 2006).

The listening audience is expanding within Japan as well. *Pipaushi* is now carried over a community FM station in Sapporo, and a station in Kobe carries the Ainu variety program. Japan's assimilation policy nearly obliterated the Ainu language. Today it is being revived, through language classes and teacher training programs. Shigeru Kayano said, "There was a time when we almost lost our language because they only taught Japanese at our schools ... now we have textbooks and dictionaries. Times have changed. People can now learn the language, if they wish to do so." He called language "the symbol of a people ... the soul. As long as we keep our language alive, our soul will continue on its shining path" (Zakoda 2006)

In 2005, the International Labor Organization (ILO) recorded a discussion between Yuuki Hasegawa, an Ainu from Japan, and Tony Khular, from the Lamkang community in Manipur, India. Hasegawa noted significant differences. The "Ainus have almost lost their language and now we want to revive it ... the Lamkang still maintain their language and culture, so we can learn from them." Government recognition differs as well. India has a number of constitutional provisions that grant rights to certain "scheduled tribes." "In Japan, there are no such rights. India has ... ratified ILO Convention No. 107," which gives rights advocates something to build on. "In Japan, ratification is too difficult and we have to find other ways to promote indigenous rights"(ILO 2005).

One of the most interesting developments is the Sámi Network Connectivity Project (SNC), started informally in Sweden in 2002, by a group called Radiosphere. In January 2004, it received funding and became a formal project. Its aim is to establish Internet communications for Sámi reindeer herders living in remote areas, to facilitate networking that takes into account the realities of people who move seasonally, according to the movements of reindeer. "This population currently does not have reliable wired, wireless or satellite communication capabilities in major areas within which they work and live ... A radical solution is therefore required" (SNC 2006). SNC has parallels in northern Canada, Australia, and elsewhere in the world where Indigenous people live in remote communities or regions with challenging climate and terrain. The founding directors are Maria Udén, Project Manager, Luleå Tekniska Universitet (Lulea University of Technology) in Sweden; Avri Doria, Project Architect, of ETRI in Daejeon, Korea; and Durga Prasad Pandey, Indian School of Mines, Dhanbad, India.

SNC's starting point is to find a "hybrid solution" using current Internet technologies that is adaptable to the needs of remote and nomadic communities. "The initial goal is to provide email, cached web access, reindeer herd tracking telemetry and basic file and data transfer services using Delay Tolerant Networking (DTN) technology under development by the Internet Research Task Force and other strategies developed by the Stockholm Open Net project." The intention is to develop facilities that are mobile, and culturally and environmentally sensitive, "enabling families to better keep together during the nomadic year cycle via arrangements such as distance learning (contemporary re-establishment of nomadism, i.e., 'post-modern nomadism')" (SNC 2006).

> From "tele-health" to internet radio, Information and Communication Technologies (ICT) have changed the way the world operates. The United Nations and the International Telecommunication Union (ITU) established the World Summit on the Information Society (WSIS) in recognition of the need for a conversation to address the potential advancements of the digital revolution and how best to bridge the digital divide between developed and developing countries (Brown 2005: 1).

As part of WSIS, the Canadian government and the Aboriginal Canada Portal and Connectivity Working Group convened the Global Forum of Indigenous Peoples and the Information Society (GFIPIS) in 2003 in Geneva, in cooperation with an array of Indigenous peoples, the UN Permanent Forum on Indigenous Issues, and other UN agencies. The "strong indigenous presence" "led to the integration of indigenous concerns in the WSIS Declaration." A follow-up meeting in Ottawa later that year considered ways to bridge the digital divide between Indigenous and other peoples. One result of the meeting was the creation of a WSIS International Indigenous Steering Committee to help assure follow-through in the next stage, including assurance of continuing Indigenous representation in the process

The steering committee comprises a male and female Indigenous representative from each of the seven regions recognized by the UN Permanent Forum on Indigenous issues; a state member from each region; and members from UN agencies, NGOs, the academic community, and the private sector. The WSIS Declaration of Principles states that, as the "Information Society" evolves, "particular attention must be given to the special situation of indigenous peoples, … the preservation of their heritage and their cultural legacy." The WSIS Plan of Action includes a commitment to developing education and training programs; implementing policies "that preserve, affirm, respect and promote diversity of cultural expression and indigenous knowledge and traditions" and supporting local "content" and resource development with an emphasis on Indigenous knowledge, traditions, and languages (Brown 2005: 1).

Indigenous peoples' efforts to work together globally are rapidly escalating. In 2008, Ainu hosted the first in a series of Indigenous Peoples' Summits, organized around the G8 summits. The result was the Nibutani Declaration, drafted by the more than 600 participants from Ainu Mosir (Hokkaido), Uchinanchu (Okinawa), the United States, Canada, Hawai'i, Guam, Australia, Bangladesh, the Philippines, Norway, Mexico, Guatemala, Nicaragua, Taiwan, and Aotearoa/New Zealand (Indigenous Peoples Summit in Ainu Mosir 2008). Here are excerpts from the Nibutani Declaration:

> We want to express our profound concern over the state of the planet … We believe that the economic growth model and modernization promoted by the G8, which suggests that we can control and dominate nature, is flawed. This dominant thinking and practice is responsible for climate change, the global food crisis, high oil prices, increasing poverty and disparity between the rich and the poor, and the elusive search for peace, the themes which the G8 nations precisely want to address in this Hokkaido Toyako Summit.

The issues and concerns expressed in the document include:

- Violations of civil, political, economic, cultural, and social rights
- militarization of Indigenous communities, arrests, and killings of Indigenous activists; use of national security and anti-terrorism laws to criminalize resistance actions

- failure to recognize "our collective identities as indigenous peoples"
- theft of intellectual property rights
- desecration and destruction of sacred and religious sites
- the shift away from production of food crops to crops for biofuels
- decreased control over, and access to, sources of subsistence
- "aggressive promotion of chemically intensive industrial agriculture"
- increased extraction of oil, gas, and minerals from Indigenous lands
- increasing loss of indigenous languages and cultures

The Declaration proposed that the G8

- implement the UN Declaration on the Rights of Indigenous Peoples; the governments of Canada, the US, and Russia adopt the Declaration and press the governments of New Zealand and Australia to follow suit
- ensure and facilitate participation of Indigenous peoples in the UN Framework Convention on Climate Change (UNFCCC); establish a Working Group on Local Adaptation and Mitigation Measures of Indigenous Peoples
- cancel funding for large hydro-electric dams; reject proposals to include nuclear energy among designated forms of "clean energy"
- promote development of small-scale, locally-controlled, renewable energy projects
- criminalize "hoarding of food by food cartels and syndicates"
- criminalize the export of banned toxic chemicals, fertilizers, and pesticides to Indigenous communities
- bring corporations from G8 countries causing environmental damage and committing human rights violations to justice and require them to compensate the affected communities
- support Indigenous campaigns against militarization, killings, mislabeling of Indigenous activists as terrorists and use of national security acts and anti-terrorism laws to curtail "our legitimate resistance against destructive projects and policies"

- support efforts to bring complaints against States that violate Indigenous rights, before UN treaty bodies, regional commissions, and human rights courts
- support cultural centers, museums, educational institutions, and programs
- stop theft and piracy of traditional Indigenous knowledge
- stop nuclear proliferation, weapons use of depleted uranium, and dumping of radioactive and other toxic waste in Indigenous territories
- support implementation of the Convention on the Elimination of Discrimination against Women in each nation
- remove US military bases from Indigenous peoples' territories; bring to justice military personnel "charged with rape and sexual assault of Indigenous women"
- urge the Japanese Government to work in concert with the Ainu community, to implement the UN Declaration on the Rights of Indigenous Peoples and make it national law, in Japan

For its own part, the group pledged to

- establish a network of Indigenous Peoples to organize and convene summits in connection with each G8 Summit; "Indigenous Peoples all over the world" will take up the responsibility to implement the UN Declaration and "enter into constructive dialogue with States, the UN … and the other intergovernmental bodies"
- develop Māori-style language nests
- establish an Indigenous Peoples Green Fund
- support the struggle "for land justice" and the "return of forests and traditional lands"

The declaration was agreed to on 4 July 2008 by representatives of Ainu (Japan), Ami (Taiwan), Sing "Olam Igorot (Philippines), Kanakana'ey Igorot (Philippines), Juma (Bangladesh), Chamoru (Guam), Hawai'i, Maori (Aotearoa/New Zealand), Uchinanchû (Japan), Yorta Yorta (Australia), The Saami Nation, Maya Kachikel (Guatemala), Miskito (Nicaragua), Nauha (Mexico), Cherokee (USA), Comanche (USA), Isleta and Taos Pueblo (USA), Jemez Pueblo (USA), Mohawk (Canada), Lakota Sioux (USA), St'at'imc (Lil'wat) (Canada), and Victoria Tauli-Corpuz,

Chairperson of the United Nations Permanent Forum on Indigenous Issues (Indigenous Peoples Summit in Ainu Mosir 2008).

Changing the Media, Using the Media for Change

From the Ainu Mosir summit to the founding of WITB, Indigenous people are changing the media, and with it, the world. Taking over the media brings new perspectives to old issues. Everett Soop did it by simultaneously and fiercely challenging both mainstream and Indigenous politics and society. John Amagoalik is widely considered to be the "father of Nunavut" for his long years of lobbying for Inuit self-determination. He has made numerous public and published statements about the persistence and longevity of Inuit language and culture. His column, "My little corner of Canada," written for *Nunatsiaq News*, offers insightful and sometimes unsettling perspectives on all manner of issues, from local and national sports to global politics. In one column, titled, "From igloos to boardrooms," he challenged the prevailing language used to refer to Inuit land claims. He protested the reference by Qallunaat (non-Inuit) to "pre-settlement traditional areas" as "camps," saying, "I don't like the word 'camps,' because we were not camping, we were living" (Amagoalik 1998: 9).

In June 2006, Aboriginal Voices Radio (AVR) began broadcasting "a mix of music, talk, news and cultural programming" from three regional studios, with additional stations joining the network shortly thereafter. Founded by the multitalented journalist, actor, editor, and activist Gary Farmer, it is Canada's first Indigenous nationwide radio network. Though is not linked, as originally planned, to the nationwide television programming of APTN, AVR parallels the purposes and enriches the programming of APTN. AVR aims to provide "a national voice" for Indigenous people (Andrews 2006: 1).

Today's Indigenous communications share some history with all mass media, having evolved from earlier ways of sending and sharing information. Spender traces today's media back to a historical turning point. "The printing press changed the course of human history. It produced an information revolution. It changed what human beings know, and how we think" (Spender 1995: 1). Before the printing press, exchange of information was limited to the religious and political elites. When the presses started rolling, there was no stopping the spread of news. The speed of Net-surfing relative to the earlier speed of, say, typing and sending a letter, is roughly equivalent to the speed of a rotary press relative to the speed of a scribe-written, hand delivered letter or manuscript. Nothing changes; everything changes. The idea of linking speed and access to the democratization of information-sharing dates back to the fifteenth century.

> More and more of the world is becoming "wired." We are entering a generation marked by globalization and ubiquitous computing … considerations and applications of social responsibility must be broader, more profound and above all effective in helping to realize a democratic and empowering technology rather than an enslaving or debilitating one (Rogerson and Fairweather 2003: S4).

New technologies do not automatically empower people or improve communications. However sophisticated they may be, they are designed and run by people, and are only as effective as the people who are involved. In fact, new technologies offer empowerment and disempowerment in nearly equal measure. The public has two main forms of access: the material derived from known and unknown sources, available on television, and the wider array of sources available on the Internet. Individuals and media organizations are using the Internet to expand public access to information. There is one major limitation—access to this new information universe is limited to those with access to computers. Canada has a strong record of providing media services to its citizens.

> We now find ourselves in the middle of the age of communication ... the millennium is upon us and it seems at times we're more isolated than ever before. TV and computers are taking us further and further away from human contact. Yet, we yearn for human interaction … Seventy-five years ago, we were all we had except for Saturday night radio. A hundred and fifty years ago, we were all we had. We must have been a hell of a lot more fascinating. We had to have been great storytellers. We certainly had a lot more time for each other ... No TV to keep us amused, or e-mail love to find a relationship. Yet, the amount of information we have at our finger tips is astounding. Access such as we have never had before...
>
> I wonder about our existence in the future and what communication skills will be required to survive...Can we live off our creativity and our ability to communicate? (Farmer 1998:6).

The following case study describes how this has empowered people in the tiniest, most remote communities, as well as in the more media-privileged cities.

Technology for People

Around the planet, Digital Radio is gaining momentum. Radio no longer comes just from small or large listening devices. It is accessed via mobile, or cell phones, digital television, and the Internet (Plunkett 2007: 9). One example of what is happening is Wapikoni Mobile—a traveling film production unit for young Indigenous people of Québec founded in 2002 by Manon Barbeau, with support from the National Film Board of Canada, launched in June 2004. Its mission is "to teach digital technology know-how to young people of the Aboriginal communities and to guide them through the universe of script-writing and production." It sends young non-Indigenous filmmakers out into Indigenous communities to recruit young Indigenous people and provide training, support, and equipment to develop "projects that reflect them." The project culminates in screenings of the films they produce (Lagueux 2006).

Farther north, the Arctic is a huge "small town." Its sparse population is distributed across an enormous region, yet its position as the globe's smallest populated circle makes the interconnectedness of separate countries, cultures, and communities exceptionally transparent. It is one reason Inuit have one of the world's most successful minority organizations, the Inuit Circumpolar Council. The Arctic's small town qualities have persisted in the computer age. In many ways, the Internet and satellite technology have picked up where town meetings left off, and in some cases, they have *become* the town meetings. Satellites carry community-to-community meetings, enabling people to convene in weather that would prohibit air travel or to bridge distances beyond the capabilities of travel budgets.

The Internet carries interpersonal, intergroup, inter-regional, and international dialogue. It helps bring Aboriginal people and Aboriginal media to each other, as well as to non-Aboriginal people. It helps to "mainstream" marginalized media—which is a mixed blessing, since "alternative" and "minority" media have in the past both lost and gained from privacy and smallness. As Gary Farmer says, there is a need for caution and for careful education about the beneficial and damaging uses of new and emerging technologies. In 1990, the Internet was the province of a few "techies." Today, it is rapidly becoming the primary medium of expression for the voices of many individuals and peoples. Of the earlier communication media, only radio continues to relay grassroots information frequently, freely, and rapidly, within communities, and from community to community. In terms of community-to-the-outside communication, the Internet is far more effective. Thus, what may have begun as carefree local expressions are becoming increasingly conscious media for sending out carefully chosen and constructed messages. The same concerns Aboriginal journalists are expressing about who controls their newspapers affect the Internet communications (which are often linked directly to the newspapers, and therefore are subject to the same struggles for power, expression, and control).

The one unarguable truth is that these sites are expanding and increasing and are likely to continue to do so in the foreseeable future. In 1998, all 58 of the communities in the Northwest Territories (including the portion which in 1999 became Nunavut) were connected to the World Wide Web. The two-year project was undertaken by a consortium of predominantly Aboriginal-owned northern businesses, Ardicom (Zellen 1998: 50). Jim Bell, editor of *Nunatsiaq News*, sees great potential for Aboriginal communities in CD-ROM, digital video discs, and other multimedia technologies, which can provide storage for important audio and visual material (Zellen 1998: 52). These technologies offer an effective way to store and transmit oral histories, materials from community and regional archives, and music and visual art—resources which are currently available or are being developed. Bell thinks the Internet offers a way of "fighting back"—a chance to "send the information the other way" (Zellen 1998:52) and an antidote to the cultural demolition that has occurred in some other media. Zellen contrasts the new technologies' culture-preserving potential with the negative effects of television described in Rosemarie Kuptana's famous speech.

The possibilities for teleconferencing and other techniques are still in the early stages of exploration. In Nunavik (Inuit northern Québec), students at Ulluriaq School in the village of Kangiqsualujjuaq are studying the violin. They played their first concert in 2003. The lessons and rehearsals were carried between Kangiqsualujjuaq and their teacher's home in Buckingham, Québec—thousands of kilometers to the south—by satellite.

> The satellite connection … uses protected bandwidth … because this connection is separate from the Internet, it allows for high-quality videoconferencing … Every week, [the school principal and the teacher] begin the violin class by linking the two schools' computers using their broadband satellite connections (Nelson 2003: 20–21).

Students and teachers can watch each other and work together individually and in groups on high-quality computer monitors. The technology is being extended to carry Inuit cultural traditions of throat singing, drum dancing, and carving from community to community, as well as carrying non-Inuit cultural instruction from south to north.

Sámi have well-established websites on which they communicate in Swedish, Norwegian, and Finnish, and also in Sámi languages, using type fonts developed by Apple. Apple also developed an Inuktitut program for use by Canadian Inuit in Nunavut and Nunavik, and Dene fonts in several Athabaskan languages for both Windows and Macintosh computers. In the polar countries, northerners are often more familiar with high-tech communications than southerners. Sophisticated computers have been commonplace in adult education and other centers for years. Satellite dishes sit on ancient rocks in tiny settlements whose inhabitants still spend much of their lives on the land.

Gary Farmer's concerns about the limitations of technology are heard widely in Aboriginal communities everywhere. However, especially in northern and remote communities, few people advocate avoiding the Internet or ignoring its potential power. Instead, the trend is toward using that power to communicate the messages and languages of the senders, clearly and carefully, from community to community and from Aboriginal communities to the outside world. By the time this information reaches you, I expect the list of Indigenous Internet sites will have doubled.

While Gary Farmer struggles to find the proper balance between human and electronic contact from his informationally advantaged, urban home base in Toronto, the founders of Nunavut—Canada's newest, Inuit-majority territory—are embracing new telecommunications technologies without hesitation. In 1996, the Nunavut Implementation Commission released a report about the role they hoped telecommunications would play in the new territory. "The road to Nunavut is along the information highway," they wrote (Nunavut Implementation Commission 1996; Bell 1996c: 16–17). The new digital network is creating jobs for Nunavut citizens (a technician for each community, regional technologists, and systems engineers), developing and linking libraries and databases, and establishing a sophisticated videoconferencing system that—in a region in which climate and distance make physical travel costly and difficult—amounts to an arm of government.

The unique conditions of northern life are ideally suited to such experimentation. While teleconferences cannot replace direct human contact, they can certainly replace a good deal of the necessary but very difficult travel, which stalls government processes and absorbs northern budgets. In Nunavut, futuristic communications systems will help maintain the Inuit way of governing, with power widely distributed and structures decentralized. The widely scattered, remote communities are able to stay in touch without the dangers and delays experienced in the past. The system allows members of the

Figure 5.3. Satellite dish at Pangnirtung, Nunavut, 1984. Photograph by Valerie Alia.

Legislative Assembly to attend meetings in their home communities, without having to leave Iqaluit (the capital), and to attend caucus meetings in Iqaluit, without having to leave their home communities.

Nunavut's citizens have greater access to each other and to government; a videoconferencing system is helping to reduce the enormous travel budget. Thanks to the well-developed, Inuit-directed broadcast industry, people in Nunavut are already comfortable with adapting television to their own needs. They were among the first Canadians (and the first Aboriginal Canadians) to take television and radio into their own hands.

In Nunavut, interactive computers provide a new medium for print and are swiftly evolving into media that transmits visual and auditory material as well. Interactive radio carries voices across great and small distances. Interactive television goes a step farther, allowing body language and facial expressions to be carried. Soon, virtual media will produce facsimiles of "real" experience, including tactile experience. None of these can replace the joys and stresses of communicating in person. But then, news media have not been face-to-face since the days when troubadours and town criers roamed the streets. Their closest descendants are probably the storytellers who still inhabit longhouses and tents, snow houses, friendship centers, and public libraries.

Indigenous People and the Information Society

Terence Turner notes the "Kayapo penchant for using video to document not only historic encounters with Brazilian state power but internal political events as well, such as meetings of chiefs or the founding of a new village" (Turner 2002: 87). *"All over the world people are looking at these videos we are making of ourselves … These videos will be seen in all countries. Tell your children and grandchildren, don't be deaf to my words, this [work] is to support our future generations, all our people … I am a Kayapo doing this work. All of you in all countries who see the pictures I make can thereby come to know our culture, my culture of which I tell you today"* (Mokuka, Kayapo leader and videomaker, translated from Kayapo by Terence Turner, in Turner 2002: 73).

In acknowledging and highlighting the power of media networking and the brilliance of Indigenous media, cultural, and political strategizing, we must also acknowledge the sometimes insurmountable obstacles. Threatened state governments, corporations, and others do not appreciate what Paul DeMain called media "guerillas," let alone wish to share "legitimate" power, privilege, and lines of communication with them. In the worst times and places, the penalty for challenging the power of oppressive states and constituencies is to be silenced. Attempts at censorship are increasingly thwarted by inventive use of new technologies and old information networks, including the oldest method of sending messengers and couriers from place to place to share the information personally. Alongside some of the most magnificent accomplishments in the dissemination of news and information in unprecedented ways, journalists are dying in unprecedented numbers. And yet, the courageous journalists continue their work.

In countries "where there is no political freedom, press freedom [is] an ambition that leads people to the loss of their liberty and, increasingly, to their deaths." Reporters sans Frontières (RSF) produces an annual index of the state of press freedom in 169 countries. Its criteria include the number of journalists killed in each country, and also "the degree of freedom" accorded journalists and news organizations, and the efforts of authorities "to guarantee respect for that freedom," including the freedom to access information on the internet. The British journalist and media critic, Roy Greenslade, says the index "importantly includes the degree of impunity enjoyed by those who are responsible for violating freedom" and "measures the level of self-censorship in each country and the ability of the media to investigate and criticise" (Greenslade 2007 [unpaginated]). Greenslade and others are frustrated and angered by the lack of attention that officials in many countries pay to the killing of journalists. "I have written often about journalists dying in Pakistan, Mexico and the Philippines. In those countries, it seems, the authorities make little, if any, effort to investigate the murders of reporters" (Greenslade 2007).

In some cases, journalists are targeted; in others, the targets are Indigenous leaders. In 2005, in the Colombian town of Caldono in the department of Cauca, police raided the home of former mayor and noted Indigenous leader, Vicente Otero, claiming that his home was harboring an "arsenal." He went into hiding. "The idea that Otero had an arsenal in his house is preposterous on every level. A lifelong pacifist, well-known political organizer and leader in the indigenous movement, Otero ... works intimately with the 'guardia indigena,' the unarmed 'indigenous guards' who ensure security throughout the region." Rather than a legitimate claim, the police action is said to be part of "an established pattern of government repression in indigenous territory" (Podur 2005 [online, unpaginated]).

The Committee to Protect Journalists (CPJ) issues an annual report on press freedom worldwide. The report for 2007 found that "64 journalists were killed this year while performing their jobs" and CPJ researchers were investigating another 22 deaths to determine whether they were also job-related; if so, the tally could rise to as much as 86. Either way, the year's death-count was seen as "unusually high," and according to CPJ, many of the journalists "were victims of targeted killings" (Rizvi 2007). The chilling fact is that worldwide, "murder remains the leading cause of work-related deaths for journalists," reflecting both the courage of journalists who know they are at risk and the degree of intimidation leveled against them. "In every region of the world, journalists who produced critical reporting or covered sensitive stories were silenced [and] media support workers are increasingly at risk." In 2007, 20 translators, "fixers," guards, and drivers were killed, along with the journalists for whom they provided support (Rizvi 2007).

While remaining targets, journalists in the field are almost never armed. Those who survive have a continuing, world-changing mission. In Paul DeMain's overarching plan, Indigenous "media guerillas" are given education, not guns. Starting in countries and regions with relative freedom, they move towards infiltrating and transcending the system, until gradually, they are able to reach not only Indigenous audiences, but the wider public. As DeMain sees it, "Our challenge is putting people into the pipeline. How do you worry about getting someone into the editorial department of the New York *Times* when you've got to get them out of High School? We've got to get them into college and keep them in college" (DeMain 2001: 132).

Towards a theory of cultural resiliency

> Cultures, ways of thinking, attachments to community, are much more resilient than many scholars of society have supposed. —(Cohen 1985:75)

My early research on identity in Nunavut and on Indigenous media, culture, and language suggested that the interference some called cultural genocide, or ethnocide, was substantial. In the mid 1990s, a decade after the research was begun, the evidence suggested a need to revise those findings, as I have learned that cultural genocide is far less pervasive, and cultural resiliency and regeneration far more substantial than I had thought. There is a metaphor that may be useful for understanding this. Medical science has shown that some types of brain damage, formerly thought to be permanent and irreparable, may be subject to recovery. For example, brain tissue is able to regenerate in ways thought impossible before. This depends on the extent and location of the damage, the physical and psychological characteristics of the individual, and the particulars of his or her environment.

In Canada, Annie Shapiro awoke from a 30-year coma. In the United States, after six years in a coma, Christa Lilly suddenly returned her mother's daily greeting, spent three days speaking, eating, and behaving normally, and then slipped back into the coma. Ms. Lilly's mother, Mrs Minnie Smith, said, "Hopefully the next time she comes out it'll be for good, to stay … I never gave up on my daughter and I never will" (Luscombe 2007: 17). "It is the fifth time that Ms Lilly has enjoyed a period of apparent consciousness since her brain was injured by [a] heart attack and stroke." One of her neurologists, Dr Randall Bjork, told the *Denver Post*: "There's really no reference point in our medical literature … Christa's brain is on the ropes, so to muster enough energy to have three days of conversation she has to be in relative hibernation for months … This is a miracle … I can't explain it. We don't have a name for this, but I call it 'cyclical awakenings'" (Luscombe 2007: 17). His comment echoes the title, and perhaps the medical observations, of Oliver Sacks' earlier book, *Awakenings* (1973/1983), and the film of the same name, which document the temporary "awakenings" of a group of people rendered catatonic by encephalitis lethargica, an early twentieth century epidemic. In the film, directed by Penny Marshall, Robin Williams plays the fictionalized Dr. Sayer who administers the then-new drug, L-Dopa to its startling effect (Marshall 1990). The tale is also taken up by Harold Pinter, in his play, *A Kind of Alaska*, which explores the experience of one patient, Deborah, who awakens from a 29-year sleep (Pinter 1982).

A 2006 report by researchers at Cambridge University "showed that patients who are apparently unconscious often have more brain function than previously suspected." Dr Pamela Klonoff, clinical director of the Barrow Neurological Institute, says that most "significant traumatic brain injuries produce a period of memory loss and ... confusion" lasting from a few minutes to several months. Brain cells that die are not replaced by new cells, but cells that are damaged but not totally destroyed sometimes recover, and sometimes cells can be trained to take over the function of cells that were lost (Curry 1998; Swiercinsky et al. 1993). I believe that in a comparable way, damaged cultures have considerable powers of regeneration, depending on the nature, location, and extent of damage.

In Nunavut and elsewhere, where I have looked at the politics of naming (Alia 2007), when threats and perceived threats to cultural continuity are removed, "underground naming"—the use of private (sometimes secret) names to retain one's culture—moves back above ground. Inuit are restoring traditional names or combining them with non-Inuit names and usage—through legal channels or just through daily use. They are naming babies in the old ways. There are comparable changes to Indigenous communications, and to the support Indigenous media provide to linguistic and cultural retention and regeneration (See Alia 2007).

Just as a person in a coma may appear to be dormant, but often has consciousness, people experiencing various stages of cultural coma may also retain consciousness. Cultures are never static. Each "separate" culture is linked to others in a substantially transcultural world. Cultural resiliency and transcultural sharing are part of the same social fabric. Survival depends on finding ways to celebrate commonalities and differences, while equalizing power. Like the intermittent, sometimes surprising and unexpected medical "awakenings," cultural rebirths, continuities, revivals, and awakenings are not linear. They may bubble up from time to time and place to place, or they may coalesce into border-crossing and sometimes global movements. In many media analyses, technological "advance" or change gets far too much attention. Technologies do not emerge, or exist, in a vacuum. If we are to understand their possibilities and limitations, we must remember that human involvement is paramount.

> Technologies are only the tools through which we carry on our relationships with nature (Manuel and Posluns 1974: 13).

According to Christensen, in 1996, following Iceland's lead, Greenland became the second country to establish a fully digital communications network (Christensen 2003: 61). He is right to stress that this represents far more than mere technological advance. The "cultural and identity affirming use of the Internet amongst Inuit" is a "continuous reality" that "binds elements such as new information technology, tradition, language, identity, history and many more important aspects into a Web of self-determination that makes up some of the core elements of identity: not necessarily to sustain things as they are but to carry on continuously and dynamically" (Christensen 2003: 12, 18). Furthermore, as Rachel Qitsualik has also pointed out, the use of new technology by Inuit (and other Indigenous peoples) is not a peculiarity, nor a sensation, nor a corruption of culture, but rather a common part of a continuous (re)shaping and integration of old and new elements in the lives of the Inuit. It is only outsiders who sometimes see this as anomalous or inauthentic. "The amount of technology in the daily lives of Inuit make them no less fully fledged Inuit than before, unless they themselves feel so" (Christensen 2003: 21).

Sneja Gunew calls the colonial, pseudo-multicultural program of voice appropriation "ventriloquizing ethnicity" or "cultural ventriloquism" (Gunew 2004: 74–75). Rita Shelton Deverell (1996: 62) offers an alternative program to ventriloquism, in which power relations are radically altered and journalists report "the view from the losing side". A distinguished Canadian broadcaster, arts journalist, actor, scholar, and writer of African American descent, she was one of the founders of Canada's multicultural faith channel, Vision TV, and served as news director and trainer for the Aboriginal Peoples Television Network (APTN) in its early years. Of course, that very shift in power constitutes a shift in "sides," in which the "losers" become "winners." Deverell's dedication to using the media to shift the power balance and assure a multiplicity of voices recalls the words of Anishinabe (Ojibwe) environmentalist, community leader, and human rights activist, Winona LaDuke:

> I believe that, in its foundational … ethical sense, democracy is about the need for public policy to fill the needs of the poorest people in this country, not the richest—the right for those who are the most silent to have a

> voice, the right for all of us to be able to participate in political discourse and dialogue (LaDuke 1998: 69).

Along with Clifford, Pratt, and others, Escobar (writing on changes to anthropology) reminds us that "space and place have moved from being *fixed co-ordinates* in the ethnographic map to realising that people … *move*. With the anthropological loss of place as a metaphor for culture, anthropology is now allowed to capture mobility, which involves *deterritorialisaton* and *reterritorialisation*" (Escobar 1994: 228; Olwig and Hastrup 1997; Christensen 2003: 27; Clifford 1997; Pratt 1992). Amin Maalouf is an eloquent celebrant of cultural multiplicity and transcultural respect. His vision of diversity (marred only by his decision to represent all of humanity with the masculine pronoun!) is that:

> Each of us should be encouraged to accept his (*sic*) own diversity, to see his identity as the sum of all his various affiliations, instead of as only one of them raised to the status of the most important, made into an instrument of exclusion and sometimes into a weapon of war. … everyone should be able to include in what he regards as his own identity a new ingredient … the sense of belonging to the human adventure as well as his own (Maalouf 1996/2000: 159, 163–164).

The New Media Nation brings together the various strands of Indigeneity, which in turn may someday be more fully incorporated—though never fused—into that "human adventure." Referring to Canada, Roth observes,

> First Peoples' self-organized media projects, the clustering of their broadcasters into policy lobby groups, and their cross-cultural programming initiatives have transformed them into new social actors who do media politics differently … bypassing completely or tiptoeing lightly past technological and social constraints (Roth 2005: 229).

I think Roth's observations apply to much of the Indigenous media world. The New Media Nation is part of a rapidly growing international media movement of Indigenous peoples. As they organize their own media "empires" (though without the kind of capital and power associated with mainstream media empires), other minorities and disadvantaged groups are developing their own media organizations and coordinating their interventions into existing media. Thanks in part to new media technology, local and regional minority voices are becoming global choruses, and reaching an ever-expanding audience. The literary scholar, Harald Gaski, thinks cultural borderlines are "where all the fun takes place." As in his own bicultural Sámi-Norwegian experience, he says, "one stands with both legs in both cultures." For Sneja Gunew, such transculturalisms are an antidote to cultural ventriloquism. Indigenous media do more than this: they serve the wider public. As Faye Ginsburg points out, Indigenous media producers and cultural activists are creating alternatives in the midst of "the growing corporate control of media," not only through innovative structures, projects, and products, but through "the social relations they are creating" (Ginsburg 2008: 303-04).

Grounded in cultural specificities and intercultural commonalities, and committed to creative, pan-Indigenous networking, and the broadest possible dissemination of information, the New Media Nation is a catalyst for identity assertion and transformation, a multidimensional international movement, and a force for positive global change.

Chronology of Key Events and Developments

11,000–10,000 BC	People begin to arrive in Fennoscandia, as glaciers are melting (Lehtola 1997: 19–20).
4000–500 BC	Archaeological evidence suggests that in the period immediately following the ice age, settlements arise throughout the regions now found in Finland, northern Norway, Sweden, and Russia's Kola Peninsula. Cliff drawings and amber artifacts show cultural production of peoples living in this region from the Neolithic period to the end of the Bronze Age (Lehtola 1997: 20).
800 BC	"Distinctive features of the Sámi culture" begin to appear (Lehtola 1997: 21).
200 BC	According to current scholarship, the Finno-Ugrian language in Fennoscandia splits into the Finnish and Sámi language groups during this period of the Bronze Age. Coastal peoples begin to farm; inland peoples continue their gathering and hunting way of life (Lehtola 1997: 20–21, 22).
100–50 BC	Sámi language is well established (Lehtola 1997: 21).
1200	Māori arrive from Polynesia, in what is now Aotearoa/New Zealand.
1323	Nöteborg Peace Treaty of 1323 defines the borders between Novgorod, Russia and Sweden, Karelia and Savo in the East (Lehtola 1997: 22).

1328	Telge Agreement strengthens Sweden's northern border and protects Sámi rights against colonization (Lehtola 1997: 23).
1595	Teasing Peace Treaty defines border between Sweden and Russia and contributes to the separation of eastern and western Sámi cultures (Lehtola 1997: 23).
1613; 1617	Treaties of Knäred and Stolbova, respectively, establish a lasting peace in Sápmi (Lehtola 1997: 23).
1600s to 1800s	Sámi reindeer herding emerges across Sápmi, as wild reindeer decrease; efforts are made to resist outside colonization. At the same time, in the latter part of this period, Sámi are sometimes both colonized and colonizers (Lehtola 1997: 26, 31–32).
1800s	Actions by the various governments result in the gradual closing of borders between regions of Sápmi, which is now divided into four parts located in Sweden, Norway, Finland, and Russia. This interferes with traditional reindeer migration routes and ends Finnish Sámi fishing on the Norwegian Arctic coast and Sámi fishing in the Deatnu River on the Norway–Finland border (Lehtola 1997: 36–37).
1828	The *Cherokee Phoenix* newspaper begins publishing in Georgia, USA (Levo-Henriksson 2007: 27).
1836	Launch of *The Aboriginal*, or *Flinders Island Chronicle* in Australia.
1855	Japanese Bakufu government brings in a policy of assimilation for Ainu people.
1861	Publication of the newspaper, *Atuagagdliutit* ('something offered for reading'), in Greenland.
1869	Establishment of Japan's Colonization Commission Office, which in 1871 forbade the practice of Ainu culture and use of the Ainu language.
1870–1914	Assimilationist "Norwegianization" policy undermines Sámi language, culture, and political rights; it coincides with the broader movement of European colonization. Parallels are seen in Sweden, Finland, and Russia (Lehtola 1997: 44–46).

1890 — Telegraph cable is installed, linking Canada with Bermuda (Miller 2003).

1898 — Founding of the first Sámi language newspaper, *Nuorttanaste* (*Eastern Star*), still published today (primarily religious).

1901 — Hokkaido Aboriginal Children's Educative Curriculum Act segregates Ainu and Japanese primary school students.

December 12 — St. John's, Newfoundland (Canada). Marconi receives the first transatlantic radio message from Poldhu, Cornwall (England) (Miller 2003).

1902 — Marconi establishes wireless telegraphy station at Glace Bay, Nova Scotia, licensed by the Canadian government (Miller 2003).

1904 — *Lapparnas egen tidning*, predecessor of *Samefolket*, begins publishing in Sweden, and is distributed across Sápmi. It carefully maintains its freedom from governments and government agencies, and claims to be the world's oldest Indigenous-controlled publication. It is published in Swedish and English (Samefolket 2006).

1904–1911 — *Sagai Muittalaegje* (*The News Reporter*), the first Sámi political newspaper is started. Founded, edited, and published for seven years by the Norwegian Sea Sámi, Anders Larsen; addresses Sámi nationalism, politics, and literature. It influences election of Sámi teacher, Isak Saba, to Norwegian Storting (National Parliament), where he serves from 1908–1912 as the only Sámi member (Lehtola 1997: 48–49).

1905 — Wireless Telegraph Act is signed in Canada.

1906 — Christmas Eve (December 24): Reginald Fessenden broadcasts a program of speech and music, transmitted to ships at sea (Miller 2003).

1906–1920 — Emergence of Sámi political activism.

1913 — Canada's Radio Telegraph Act is amended to include voice transmission as well as Morse Code.

1910–1913; 1922–1925 — Daniel Mortenson publishes *Waren Sardne* (*Fell Con-*

	versations) to address legal concerns of Sámi reindeer herders.
1917	February 6: Sámi from all regions meet in Trondheim, Norway; now commemorated by the Sámi national day. First general conference of Sámi, Trondheim marks "the birth of modern Sámi politics" (Pennanen 1998: 5).
1918	More than 200 delegates attend four-day meeting in Östersund, Sweden; Sámi publishing is born. *Samefolket* begins publishing.
1919	Founding of *Samefolkets Egen Ridning* (which becomes *Samefolket*, still published today); focused on cultural, social, and political developments.
	World's first broadcasting station, the Marconi station, XWA, goes on the air in Montreal; a year later, it is renamed CFCF (Canada's First).
1922	Canada now has 39 commercial radio stations operating and 66 stations licensed.
1923	Canada's CN Railway begins broadcasting radio service on trains.
1925	First radio transmission in Greenland (telegram) precedes the start of radio broadcasts.
1928–1929	Canada's Royal Commission, under Sir John Aird, issues 'The Aird Report,' on the future of broadcasting, which recommends establishing a nationwide, national radio system (Miller 2003).
1930s	First North American Indigenous broadcasts begin in Alaska
1932	Passage of the Canadian Radio Broadcasting Act.
	Launch of *Micmac News* in Nova Scotia, Canada.
1933	One-hour daily, national radio broadcasts begin in Canada.
	Launch of first Pacific Island radio service in Papua New Guinea.
1934	Launch of *Sápmelas*, first paper published in Sámi, now a monthly magazine.
1935	Radio service launched in Fiji.

1936	First radio program broadcast in Sámi is transmitted from the Polmak church in Norway.
1937	The Canadian Broadcasting Corporation (CBC), Canada's national broadcaster, opens its first station near Montreal.
1938	*Abo Call: The Voice of Aborigines* begins publishing in Australia.
	Torres Strait Islanders begin wireless transmissions.
1940–1945	Sámi are drawn into the Second World War and fight (sometimes against each other). Forced to evacuate, Sámi in Norway hide and find other ways to avoid or protest. German troops burn the Province of Lapland in Finland and northern Norwegian provinces.
1945	Canada joins the United Nations.
1946	NRK, Norwegian radio, begins regular Sámi language broadcasts, 20 minutes once a week, from Romsa (Tromsö). Two years later it moves to Ñáhcesuolu.
	The Native Brotherhood of British Columbia, Canada, begins publishing *The Native Voice*.
1947	Regular Sámi radio broadcasts begin at Oulu, Finland. Start of NRK's regular Sámi language broadcasts over YLE (YLE website 2006).
1949	Karen launch *Radio Kawthoolei* in Burma; it continues broadcasting intermittently through the 1980s (Brooten 2008).
1950s to 1960s	Rise of Sámi boarding (residential) schools, which affect family, culture, and community.
1953	First Sámi Conference is held in Johkamohkki, Sweden, though Sámi participants are in the minority; Swedish Sámi radio programming begins at Johkamohkki (Lehtola 1997: 60).
1956	Founding of *Ságat* (*The News*) as monthly and later, bi-monthly, Sámi newspaper.
1958	The US Atomic Energy Commission (AEC) develops plan to use nuclear explosives to excavate a harbor at Cape Thompson, Alaska.
	CBC Northern Service (radio) is launched in Canada.

1959	Kathrin Johnsen, "Mother of Sámi radio," begins 30-year career as head of Norwegian Sámi radio.
1960	CBC Northern Service broadcasts its first Aboriginal language radio program in the Inuit language of Inuktitut, via shortwave from Montreal.
	Also in Canada, the Federation of Saskatchewan Indians begins publishing *The Indian Outlook*.
1960s	"Sámi Renaissance" marks the start of Sámi television in Sweden, Norway, and (to a lesser degree) Finland.
1961	Inuit in the Eastern Canadian Arctic (now Nunavut) begin radio broadcasts.
1962	*Tundra Times* begins publishing in Alaska, in part as a vehicle for challenging the AEC's nuclear development plan.
1967	CBC TV (Canadian Broadcasting Corporation television) arrives in Canada's North; "Frontier Package" brings radio to 17 western Arctic communities.
	Chukchi-language television and radio broadcasts and Eskimo Yupik-language radio broadcasts begin, in Chukotka, USSR/Arctic Russia (Diatchkova 2008: 218).
1968	In Canada, the Alberta Native Communications Society launches *Native People*.
	Founding of USET intertribal alliance in the United States.
1968–1982	The Áltá Conflict, 10-year protest against environmental and cultural effects of hydro-electric projects in Norway and Finland, occurs.
1969	Launch of Anik satellite system in northern Canada opens new pathways to Indigenous broadcasting.
1970s	A group of Native American journalists in Washington, DC form the American Indian Press Association, which encourages tribal communities and groups to sponsor news publications, many of them for the first time; it later becomes the Native American Journalists' Association (NAJA) (Levo-Henriksson 2007: 24).

	American Indian Theatre Ensemble (later the Native American Theatre Ensemble) is founded in New York City (Levo-Henriksson 2007: 24).
	Aboriginal communications societies begin operating in Canada.
	Revival of arts; establishment of the Sámi Duodji trademark carried by Sámi artworks; 13 Sámi associations are established.
1970s to 1980s	Emergence of Sámi music and recording industry includes revival of traditional *yoik*.
1971	In the United States, the Alaska Native claims Settlement Act (ANCSA) begins a new era in Indigenous-government relations, and the first Native-American-owned station in the US, KYUK-FM, is launched at Bethel, Alaska.
	In Canada, the Northern Pilot Project (radio) in Keewatin (NWT) and northern Ontario receives federal funding.
1972	First Indigenous radio program on community radio airs in Adelaide, Australia.
1973	Launch of CBC northern television service in Canada.
	Launch of KYUK-TV, television service of KYUK in Bethel, Alaska, and *Tundra Drum* newspaper, Alaska, USA.
	Finland creates the Sámi Delegation (Parliament), which becomes the model for the Sámediggi in Norway and Sweden (Lehtola 1997: 70).
1974	Start of *Wawatay News* by the Wawatay Communications Society in northern Ontario, Canada.
1975	In Canada, launch of Nunatsiakmiut Community Television Society Frobisher Bay (now Iqaluit, capital of Nunavut) and CBC Northern Services begins broadcasting via Anik satellite, to communities in the Northwest Territories, Canadian western Arctic.
	George Manuel founds the World Council of Indigenous Peoples.

1977	Labrador Inuit file land claim asserting rights to 72,500 square kilometers of northern Labrador (CBC July 02, 2004).
	Founding of the Inuit Circumpolar Conference (now Inuit Circumpolar Council), first international gathering of Inuit in Barrow, Alaska (ICC is formally constituted in 1980). Theme is "under four flags." It takes 12 years for the Russians to come in person; in the meantime, a Soviet flag is placed in front of an empty chair at each assembly.
1978	Launch in Australia of *Junga Yimi*, newspaper published by the Warlpiri Media Association.
	Project Inukshuk, start of Inuit TV broadcasts over Anik B satellite in Northern Canada.
	Conway Jocks founds CKRK-FM Mohawk radio station at Kahnawake (Mohawk Territory, Québec, Canada).
1979	Greenland Home Rule Act comes into effect.
1980	Founding of the Central Australian Aboriginal Media Association (CAAMA).
	KNR, Greenland national radio, comes under control of Home Rule government.
1980s	Australian Broadcasting Corporation (ABC) begins supporting Indigenous radio broadcasts to rural areas.
	A group of Native American filmmakers launch the Native American Public Broadcasting Consortium (NAPBC) in Lincoln, Nebraska, "to support and encourage Native work in television, video, and motion pictures" (Levo-Henriksson 2007: 24).
	Growth of Indigenous media in Mexico.
1980–1985	Central Australian Aboriginal Media Association (CAAMA) and National Aboriginal and Islander Broadcasting Association (NAIBA) lobby for Indigenous media access (Molnar and Meadows 2001: 213).
1981	In Sápmi, Norway demolishes Áltá demonstrators' camps and continues developing a power plant, which opens in 1987. Positive outcomes include renewed political organization and 'cultural awakening'; forg-

	ing of new political relationships, establishment of Sámediggi (Sámi Parliament) in Norway.
	In Canada, Founding of Inuit Broadcasting Corporation (IBC), world's first Aboriginal television network.
1983	Launch of Northern Broadcasting Policy and Northern Native Broadcast Access Program (NNBAP) in Canada.
	ICC is granted NGO status at the UN.
	Government of Mexico's Instituto Nacional Indigenista (National Indigenous Institute) publishes the booklet, *Toward an Indigenous Video* (Salazar and Cóordova 2008: 44).
1984	Sámi Radio gets its own Broadcasting House in Karasjok, Norway (NRK website).
	Pivotal task force report, *Out of the Silent Land*, is released by the Australian Broadcasting Tribunal.
1985	Founding of *Black Nation* newspaper in Brisbane, Australia.
	Launch of Australia's AUSSAT K-1 satellites; establishment of the first CAAMA station, SKIN-FM at Alice Springs.
	Consejo Latinoamericano de Cine y Comunicación de los Pueblos Indigenas (Latin American Council of Indigenous Film and Communication, CLACPI) organizes its first festival in Mexico City (Salazar and Cóordova 2008: 45).
1986	Founding of Canada's National Aboriginal Communications Society (NACS).
	Windspeaker begins publishing in Canada.
	Official adoption of the Sámi flag.
	Fallout from Chernobyl endangers reindeer, vegetation, air, and water in Sápmi.
	Waitangi Tribunal heralds beginning of new relations between Māori and government in Aotearoa/New Zealand.

1987	Mikhail Gorbachev speaks at Murmansk, advocating establishment of nuclear weapons free zone in the North. ICC is invited to attend.
	Anaras, a magazine published in Inari Sámi language, begins publishing in Finland.
	Ainu participate in the UN International Indigenous People's Conference for the first time.
	Start of BRACS (Broadcasting Remote Aboriginal Community Scheme) in Australia.
	Launch of PACNEWS brings regional news to the Pacific countries.
1988	Native Communications Society (Canada) produces first Aboriginal national television documentary, "Sharing a Dream" by Michael Mitchell (Mohawk).
	ICC receives Global 500 Award from the United Nations Environmental Programme (UNEP). In all of its assemblies, ICC has called for an end to nuclear testing and stationing of planes, ships, and nuclear submarines in the Arctic, and has formally declared the Arctic to be a nuclear weapons-free zone.
	Imparja Television begins broadcasting in Australia; BRACS-delivered television reaches the outer Torres Strait Islands.
1989	Indigenous Russians attend the ICC assembly in Sisimiut, Greenland, and are granted associate membership. It is Greenland's first contact with Russian Inuit since Knud Rasmussen's visit in 1924, and the first contact between Alaskan Inuit and their Russian families since 1948.
	Mikhail Gorbachev formally recognizes Association of Indigenous Numerically Small Peoples of the Russian Federation North, Siberia, and the Far East (now RAIPON).
1990	NRK launches Sámi television service.
	Founding of the American Indian Producers Group, consisting of video artists, filmmakers, directors, writers, and producers (Levo-Henriksson 2007: 25).
	Sámi language act is passed, in Norway.

1991	The Indigenous newspaper, *Koori Mail*, begins publishing in Lismore, Australia.
	Canada's Broadcasting Act of 1991 recognizes that Indigenous broadcasting is "an intrinsic part of the Canadian broadcasting system"; The Canadian Radio-television and Telecommunications Commission (CRTC) grants a license to Television Northern Canada (TVNC).
1992	Russian (Soviet) Inuit are granted full membership in ICC.
	Founding of the National Indigenous Media Association of Australia (NIMAA).
	Television Northern Canada (TVNC) begins broadcasting.
	Launch of Bush Radio in South Africa.
1993	Sámi national flag is flown 'on the first Sámi National Day, February 6' (Pennanen).
1994	Launch of the *Yamaji News* in Western Australia.
	Founding of AIROS, The American Indian Radio on Satellite distribution network, headquartered in Lincoln, Nebraska (USA).
1995	Law on Cultural Self-Government is passed in Finland.
1996	Launch of the National Indigenous Radio Service in Australia.
	Launch of *Saskatchewan Sage* first Nation newspaper (Canada).
	National Film Board of Canada (NFB) launches its Aboriginal Filmmaking programme.
	Founding of Sámediggi in Finland.
1997	Adoption of Sámi curriculum in Norwegian schools.
	Japan passes Law for Promotion of Ainu Culture.
1997–1998	Canadian Television Fund commits funding for Aboriginal Language productions.
1998	Australia adopts the ATSIC Indigenous Broadcasting Policy.

	Signing of the Nunavut Final Agreement; creation of Nunavut Territory. Two separate agreements: Nunavut Land Claims Agreement and the Nunavut Political Accord establishing the government of Nunavut to be formally launched in 1999.
1999	Launch of Aboriginal Peoples Television Network (APTN) in Canada, evolved from TVNC.
	New Inuit-majority territory of Nunavut becomes official in Canada.
	Launch of Sámi Radio digital network (DAB) with transmitters at Kautokeino and Karasjok, Norway (Marshall 2006).
	Sámi Radio stations across Sápmi set up a joint Internet news site.
2000	Launch of Hopi Radio in Arizona, USA.
2001	In Japan, Ainu Radio—FM-Nibutani—is launched.
2002	Launch of Ajjiit in Canada; begins lobbying for recognition of industry and creation of a Nunavut Film Commission.
	Aang Serian Drum video project founded in Tanzania.
	Founding of *National Indigenous Times,* Canberra, Australia.
2002–2003	Ajjiit participate in working group to develop a Nunavut Film Policy; the policy is adopted by the Canadian Cabinet.
2003	Navajo Nation brings wireless internet to the reservation (Cullen 2005: 32).
	Sámi Radio "sets up an editorial office for indigenous peoples to strengthen the coverage of indigenous affairs in its own broadcasts and in NRK's other programmes" (Radio Hele Norge 1999, unpaginated).
2004	Ajjiit organizes first Nunavut Film Festival with premiere Tour of *The Snowalker.*
	Ajjiit hosts expanded Film Celebration, workshops, Arctic location scouting tour; launches industry awards (Canada).

Labrador Inuit ratify the last land claim agreement for Inuit in Canada, May 25. Agreement provides self-government over Nunatsiavut (Our beautiful land), "a region larger than Ireland," giving about 5,300 Inuit and Kablunângajuit ("people of mixed Labrador Inuit and European ancestry") the right to pass their own laws and control programs of health, education and justice (CBC 2 July 2004).

2005 Taiwan launches ITV, Indigenous Television Network.

Start of Guatemala Radio Project.

2006 Native Voice One (NVI) replaces AIROS as the main distributor of Indigenous programming in the United States (Levo-Henriksson 2007: 28).

Ta-Oi Radio launched in Lao People's Democratic Republic.

2007 Australian government issues formal apology to "stolen generation" and Indigenous people in general, for the abuses of residential schools and related policies.

Start of major new community radio project in India.

Australia launches NITV, National Indigenous Television Service.

Māori host founding conference of WITB, World Indigenous Television Broadcasting.

2008 Canadian government issues formal apology to Indigenous people for the abuses perpetrated by residential schools and related policies. In an unprecedented move brought about by lobbying of Indigenous and other leaders, Indigenous leaders are permitted to publicly respond, on the floor of Canada's House of Commons.

Japan formally recognizes Ainu rights.

APPENDIX

Native News Networks of Canada (NNNC)

Statement of Principles

Journalists can change or influence the thinking of those who are mere bystanders or news followers...How can the public judge, if they never have a chance to read? First of all, you need to hire Native writers. Let them write the stories they feel are important and let readers decide if this is what they've been looking for all these years.
—Bud White Eye, founder of NNNC

***

The human being has been given the gift to make choices, and...guidelines, or what we call original instructions. This does not represent an advantage for the human being but rather a responsibility.
—Oren Lyons (1986)

Native News Network of Canada (NNNC) is dedicated to the gathering and distributing of print and broadcast news, features, reviews, and opinion by, about and for First Peoples. Its purpose is to inform all people and enable them to make judgments on the issues of the day, and to this end, to give expression to the interests of people who are underrepresented in "mainstream" news media.

The ethical practice of journalism is paramount. An ethical journalism is a journalism of courage and conscience. It must reflect the diversity of the society, in terms of both hiring of journalists and representing people in news media. To carry out its purpose, NNNC has adopted the following principles:

1. Journalism should be fair, accurate, honest, conscientious and responsible.

2. Journalists and media outlets should be free from government or other outside interference; journalists must be free to discuss, question, or challenge private or public actions, positions or statements.

3. Journalism should

Encourage creativity in writing and broadcasting, and foster the different, authentic voices of Aboriginal people;

Promote the public's right to know;

Critically examine the conduct of those in the public and private sectors;

Expose any abuse of the public trust, evidence of wrongdoing or misuse of power;

Advocate reform or innovation whenever need, in the public interest;

Avoid unfair bias, distortion, or sensationalism in written and broadcast text and in printed or broadcast images;

Clearly distinguish editorials and opinion from reporting;

Present information in context;

Present information without irrelevant reference to gender, culture, color, "race," sexual preference, religious belief, marital status, physical or mental disability;

Treat racial, ethnic or other derogatory terms as obscenities, used in quoted material only when essential to a story;

Avoid photography or art work which fosters racial, ethnic, gender or other stereotypes;

Respect individuals' dignity and right to privacy; avoid unnecessary intrusion into private grief;

Except in extreme cases in which information vital to the public interest cannot be obtained in any other way, obtain information in a straightforward manner, without misrepresenting the journalist's identity;

except in rare cases in which information vital to the public interest cannot be obtained in any other way, tape-record, videotape, or photograph an interview only with the interviewee's knowledge and permission;

Avoid identifying the names or addresses of individuals whose safety might be jeopardized;

Preserve NNNC's independence from the vested interests of any particular individual, organization, institution or community;

Encourage thoughtful criticism of the news media as well as the society at large;

4. We recognize that journalists are citizens. An NNNC journalist should

be free to be active in the community, perform work for religious, cultural, social or civic organizations and pursue other activities of commitment or conscience, provided these activities do not distort the quality of his or her coverage;

Disclose any involvement in outside organizations, political or other activities, to the NNNC editors and the public;

Avoid covering stories in which he or she has a conflict of interest (for example, most stories concerning close relatives or friends);

Disclose to sources, if he or she is doing freelance work for a media outlet other than NNNC (for example, if an interview will be used for a story submitted to another news organization);

Avoid plagiarizing, by carefully quoting from or attributing information to other sources, and by assuring that his or her name is only used to identify the journalist's own work;

Honor pledges of confidentiality, which should be made with great care, and only when necessary to serve the public's need for information;

Make respect the watchword of journalistic practice – respect for subjects and sources, for other journalists, for the public we serve, and especially for the First Nations which are the backbone and the raison d'être of NNNC.

Valerie Alia
Professor of Journalism Ethics

Dan Smoke
President, Native News Network of Canada
Treasurer, Native Journalists' Association

April 1997

FILMOGRAPHY

Indigenous Films and Videos

This is a partial listing of some of the many films by Indigenous film and videomakers, with a few notable examples of works by non-Indigenous directors, which feature Indigenous people and subjects. Many of these are excellent classroom and community education resources. It is worth noting that a large number of the filmmakers are women. The list is growing daily, as the New Media Nation gathers force. Readers are also directed to the list of Indigenous Media Resources for organizations and media outlets, which can provide additional audiovisual materials.

Aboriginal Peoples Television Network (APTN) Inaugural Live Broadcast. 1999. September 1, Canada. English, with French and several Indigenous languages.

Aboriginal Peoples Television Network (APTN) Promotional Video. 1998. Canada.

Acts of Defiance. 1990. Director: Alec G. MacLeod. Canada. (Non-indigenous director; subject: the "Oka Crisis"—government and Mohawk people in Québec; makes an interesting comparison with Alanis Obamsawin's *Kanehsatake* film).

African Underground: Democracy in Dakar. 2007. Nomadic Wax & Sol Productions. US/Senegal. Wolof and French with English subtitles. Documentary about the impact of hip-hop music and its creators on Senegalese politics.

As Old As My Tongue: The Myth and Life of Bi Kidude. 2007. Director: Andy Jones. UK/Tanzania. Swahili with English subtitles. Award-winning documentary about the great Tanzanian singer, Bi

Kidude, who is said to be close to 100 years old, and is still giving public performances.

Atanarjuat (The Fast Runner). 2000. Director: Zacharias Kunuk. Isuma Productions. Canada. Inuktitut with English and French subtitles. The first all-Inuktitut feature, filmed near Igloolik in the Canadian arctic, has won major international prizes.

Babacueria. 1981. Aboriginal Television, Australian Broadcasting Corporation (ABC). Australia. Wonderful mockumentary featuring a turnabout look at cross-cultural encounters between Indigenous and European people in Australia. In this 'encounter,' Indigenous people are the stars and call the shots. Excellent teaching film.

The Ballroom (Chega de saudade). 2007. Director: Laís Bodanzky. Brazil. Portuguese with English subtitles. Five interwoven stories of sambas, lives, and dreams, set in an aging São Paolo dance hall. Bodanzky began making films in 1994 and made her first feature, *Brainstorm,* in 2000.

Barakat! (Enough!) 2006. Director: Djamila Sahraoui. Algeria/France. French and Arabic with English subtitles.Feature film about two women in 1990s Algeria.

Barking Water. 2008. Director: Sterlin Harjo. USA. Quest/road movie about love and self discovery.

Black Business. 2007. Director: Osvalde Lewat. Cameroon/France. French and Bamiléké with English subtitles. Documentary about government-sponsored 'enforcement' in the Republic of Cameroon, which resulted in the 'disappearance,' in 2000, of more than 1,000 people.

Bleeding Rose. 2007. Directors: Chucks Mordi and Kingsley Kerry. Nigeria/UK. English. Named Best Nigerian Feature Film at the 2007 Lagos International Film Festival. Dramatic feature about a botany professor's obsessive search for a healing plant.

Bongoland II: There is no place like home. 2008. Director: Josiah Kibira Tanzania. Digibeta and Swahili with English subtitles. Sequel to Kibira's 2003 film, *Bongoland,* a satirical tale of a Tanzanian businessman's adventures in Minnesota. In the sequel, he returns to Tanzania (fictionalized as "Bongoland") and tries (unsuccessfully) to institute US ways of management.

Bran Nue Dae (Brand new day). 2009. Director: Rachel Perkins. Australia. Musical comedy about an Indigenous teenager in 1960s

Broome. Starring Phillip Rocky McKenzie and Ernie Dingo. Based on Australia's first Indigenous musical, by Jimmy Chi.

Burwa dii ebo. 2008. Directors: Vero Bollow and the Igar Yala Collective (a cooperative of indigenous Kuna youth). Panama. Spanish with English subtitles.

*Buud Yam.*1997. Director: Gaston Kaboré. Burkina Faso Moré with English subtitles. Award-winning sequel to the 1982 feature, *Wend Kuuni*, about a sorceress's son.

The Chant of Jimmie Blacksmith. 1977. Director: Fred Schepisi. Feature film. Australia. Story of a "Half-caste" Indigenous man raised by a Methodist minister; based on the novel by Thomas Keneally.

Daratt (Dry Season). 2006. Director: Mahamat-Saleh Haroun. Chad. Arabic and French with English subtitles. Parable set in a bakery, in which former enemies (a teenager and his older mentor), in the process of baking bread together, learn ways of ending the cycle of violence.

Dances with Wolves. 1990. Director: Jake Eberts. With Kevin Costner, Graham Greene, and Tantoo Cardinal. USA. Feature film that some considered a breakthrough in Hollywood representations of Native Americans in US history. Notable for its use of Indigenous actors (though not all Lakota), attention to cultural details and Lakota language (though some have criticized its accuracy, as in the case of actors'/characters' use of gender inappropriate forms of speech).

Dersu Uzala. 1975. Director: Akira Kurosawa. USSR/Japan. Powerful portrayal of the encounter, in Siberia, between a Soviet scientist and an Indigenous man, and the ways in which they change each other's lives.

Different Lenses. 1996. Produced by KCTS Television, Seattle. USA. Television documentary on the brothers, Edward and Asahel Curtis, both photographers, and the different ways in which they depicted Native Americans/First Nations people and Eskimo/Inuit.

The 8th Fire. 2008. Directors: Cindy Pickard and Anne Pickard. USA.

Portrays an Anishnabe prophecy and the experience of Dave Courchene Jr., an elder and spiritual adviser who abandons his job for a worldwide expedition to promote peace.

Ezra. 2007. Director: Newton Aduaka. Nigeria. English with French subtitles. Based on the filmmaker's personal experiences, the film follows a seven year-old boy who is kidnapped and trained as a child soldier and later faces a 'Truth and Reconciliation Commission'.

Faat Kiné. 2001. Director: Ousmane Sembene. Senegal. French and Wolof with English subtitles. Portrait of a businesswoman who successfully owns and runs a gas station while lone-parenting two children in patriarchal Senegal. The first film in a planned trilogy celebrating the strength of African women. Sembene died in 2007, after completing the second film, *Moolaadé.*

Four Sheets to the Wind. 2009. Director: Sterlin Harjo. USA. Harjo is Native American, Seminole/Creek; this is his first feature film.

Grand Avenue. 1996. Director: Daniel Sackheim. Teleplay: Greg Sarris (Miwok/Pomo), based on his book of interlocking stories. Originally a pilot for a television series. Excellent film about three families and their experiences moving between 'reservation' and city life. The fine cast includes Sheila Tousey, Irene Bedard, Tantoo Cardinal, and A. Martinez.

Grey Owl. 2000. Director: Richard Attenborough. UK. Fictional treatment of the life of Archie Belaney, the British imposter and groundbreaking environmentalist who masqueraded as a First Nations man. The film would be more successful had it explored Grey Owl's life and work in all of its complexity.

Hands of History: Four Aboriginal Women Artists. 1994. Director: Loretta Todd. Produced by the National Film Board of Canada. Canada.

Il va pleuvoir sur Conakry (Clouds over Conakry). 2007. Director: Cheick Fantamady Camara. Guinea. French and Malinke with English subtitles. Challenging feature film about a lovestruck newspaper cartoonist in Guinea's capital, Conakry, which looks at religious extremism, political corruption, and honour killings.

Incident at Restigouche. 1984. Writer-Director: Alanis Obomsawin. Canada. Documentary of the 1981 Québec Provincial Police raid on members of the Micmac Nation at Restigouche, sparked by challenges to fishing rights by commercial and sport fishers.

The Journals of Knud Rassmussen. 2006. Director: Zacharias Kunuk. Isuma Productions. Canada.

Kabloonak. 1994. Director: C. Massot. France/Canada. Fictionalized account of Robert Flaherty's experiences in the Canadian North, and the making of his film, *Nanook of the North.*

Kanehsatake: 270 Years of Resistance. 1993. Writer-Director: Alanis Obomsawin. Produced by the National Film Board of Canada. Canada. Important documentary about the armed standoff, known as the "Oka Crisis," between the Mohawk Kanehsatake First Nation and Québec police. The director is Abenaki; one of the most respected Indigenous filmmakers both in Canada and internationally, she has had a parallel career as a singer-songwriter.

Lakota Woman: Siege at Wounded Knee. 1994. Director: Frank Pierson. USA. With Irene Bedard, Floyd Red Crow Westerman, and Tantoo Cardinal. Based on the conflict between Native Americans and the US government, the film shows the important role of radio broadcasts in maintaining community morale and solidarity.

The Learning Path. 1991. Director: Loretta Todd. Canada. Produced by the National Film Board, TV Ontario, and Tamarack Productions. Excellent documentary by one of Canada's most important Indigenous filmmakers.

Let my Whakapapa Speak. 2008. Director: Tainui Stephens. New Zealand. Portrait of Iritana Te Rangi Tawhiwhirangi. In post-war Aotearoa/New Zealand, she was one of the leaders of the Kohanga Reo Maori language revitalization movement, which has influenced Indigenous language teaching not only in Aotearoa but throughout the world.

Little Caughnawaga: To Brooklyn and Back. 2007. Writer-Director: Reaghan Tarbell. USA. Mohawk filmmaker Reaghan Tarbell traces her family's connection to Brooklyn, New York, where for more than fifty years, the Mohawks, of Kahnawake, Quebec (Canada) occupied a ten-square-block area of Brooklyn, known as "Little Caughnawaga." The men were skilled ironworkers who came to New York for work and often brought their families with them. Much has been written and filmed about the men; Tarbell's film explores the women's experience.

The Lone Ranger: Message to Fort Apache. 1954. USA. Television series. A particularly interesting episode, useful for discussing Hollywood representations of Native Americans.

Love has no Language. 2009. Director: Ken Khan. New Zealand. Cross-cultural love story of a Maori man and a woman who has immigrated to New Zealand from India. Starring Ben Mitchell and Celine Jaitley.

The Piano. 1993. Director: Jane Campion. (Representation of Māori).

Picturing a People. 1997. Director: Carol Geddes (Tlingit). Documentary film. National Film Board. Canada. First Nations filmmaker chronicles the life of a Tlingit photographer from her home community of Teslin, Yukon. Interesting multilayered portrait by an image-maker whose subject is an image-maker.

Pow Wow Highway. 1997. Director: Jonathan Wacks; screenplay by David Seals (based on his novel). Feature film. USA.

Rabbit Proof Fence/Long Walk Home. 2002. Director: Phillip Noyce. Based on the book by Doris Pilkington (Nugi Garimara), *Follow the Rabbit-Proof Fence* (Pilkington 2002).

Rain of the Children. 2008. Director: Vincent Ward New Zealand. Docudrama about Puhi, a Tuhoe woman first portrayed in Ward's 1978 film, *In Spring One Plants Alone.* At that time, Puhi was eighty and he was a twenty-one-year old art student seeking knowledge of her traditional way of life. Features Rena Owen as the older Puhi, with Miriama Rangi and Temuera Morrison.

El Regalo de la Pachamama. 2008. Director: Toshifumi Matsushita. Bolivia. Quechua with English subtitles. Coming-of-age tale of a thirteen-year-old boy who lives hear the salt lake, Uyuni. As he travels the "Salt Trail" with his father, exchanging salt for other Andean resources, he learns more about Quechua culture and experiences his first love.

Richard Cardinal: Cry from a Diary of a Métis Child. 1986. Writer-Director: Alanis Obomsawin. Canada. A short but highly influential and important documentary of Cardinal's life and the failure of social services, based on his diary. Along with efforts by his final foster parents, the film led to changes in child welfare administration and policy.

Rocks at Whiskey Trench. 2000. Writer-Director-Narrator: Alanis Obomsawin. Feature-length documentary. Canada. Fourth film in Obomsawin's series on the 1990 Oka/Kanehsatake 'Crisis' focuses on the history of Kahnawake and outside appropriation Mohawk lands.

Rua. 2009. Director: Garth Wateneh. New Zealand. A young boy's experience of his mother's death.

Spudwrench: Kahnawake Man. 1997. Writer-Director: Alanis Obomsawin. Canada. During the 1990 Kanehsatake/Oka crisis, the traditional warrior, Randy Horne, a steel worker from Kahnawake, was known as Spudwrench. The documentary uses his story to explore the wider story of Mohawks who have long been known for their steel work on bridges and skyscrapers, in New York and elsewhere in the world.

The Story of Joe and Elise. 1995. *Man Alive* series, CBC Television. Canada. Interesting but problematic television documentary about the killing of Elise Attagutaluk by her husband, Joe. The film's focus on Joe does serious injustice to Elise's life and the important work she did, including founding a women's shelter in her home community of Igloolik.[1]

Taking the Waewae Express. 2008. Directors: Andrea Bosshard and Shane Loader. New Zealand. The directors' first feature. The cast includes Matariki Whatarau and Rangimoana Taylor.

Ten Canoes. 2006. Director and Writer: Rolf de Heer. Co-director: Peter Djigirr. Written and produced in consultation with the people of Ramingining. The first feature film in Australia with all of the dialogue in an Aboriginal language, though with English narration. Starring David Gulpilil (Yolngu) as the storyteller. Set in his country and based on his traditional stories.

Under the Southern Cross. 1927. New Zealand. Early silent film, starring Maori actress,Witarina Harris, who died in 2007, aged 101.

Voice of the West: M. Scott Momaday. 1993. Producer: Jean Walkinshaw. KCTS TV, Seattle (portrait of distinguished Native American writer). Television documentary. USA.

Waban-Aki: People from Where the Sun Rises. 2006. Writer-Director: Alanis Obomsawin. Feature film. Canada. Having spent several decades filming people in other First Nations, Obomsawin has returned to her home village of Odanak, famous for its basket-makers, where she was raised. Feature film addressing cultural survival.

Walkabout. 1970. Director: Nicolas Roeg. Australia. Feature, about the encounter between two young, non-Indigenous Australians and a young Indigenous man in the outback. Feature film.

Whale Rider. 2003. New Zealand. Director: Niki Caro (Māori). Writers: Niki Caro, Witi Ihimaera. Starring Keisha Castle-Hughes and Rawiri Paratene. Visually beautiful, moving, and sensitively acted by a Māori cast. A Māori colleague praised the film's 'authenticity', but Jennifer Gauthier (a Pākehā communications scholar) criticizes it for having a Pākehā director, producer, and crew, and for failing to "embrace a Maori aesthetic" (Gauthier 2008).

The Wind and the Water. 2008. Writer-Directors: Vero Bollow and the Igar Yala Collective. Panama. Panama's first feature-length narrative film portrays turning-points in the lives of two Kuna teenagers, Machi, raised in Panama´s San Blas islands, and Rosy, who has only known urban life. Based on the Collective members' own experiences.

Yeelen (Brightness). 1987. Director: Souleymane Cissé. Mali/Burkina Faso/France/West Germany. Bambara with English subtitles. Celebrated African film about father-son conflict in 13th century Mali, with outstanding cinematography.

Zan Boko. 1988. Director: Gaston Kaboré. Burkina Faso Moré with English subtitles. Explores the experience of traditional village life and rural economies, in the midst of industrialization and urbanization.

Indigenous Networks and Media Organizations

On- and Off-line Resources

Rights and Advocacy Organizations

Aboriginal Australia. Australian Institute of Aboriginal and Torres Strait Islander Studies www.aiatsis.gov.au

Aboriginal Canada http://www.aboriginalcanada.gc.ca/international.html

Abya Yala Net (Kuna peoples of Panama and Colombia) http://abyayala.nativeweb.org/ Ecuador site: http://abyayala.nativeweb.org/ecuador/) Links to numerous sites, including: *La Confederación de Nacionalidades Indígenas del Ecuador (CONAIE); Confederación de Pueblos de la Nacionalidad Kichua del Ecuador (ECUARUNARI); Confederación de Nacionalidades Indígenas de la Amazonia Ecuatoriana (CONFENIAE); Federación de Comunas Unión de Nativos de la Amazonía Ecuatoriana (FCUNAE); Federación Ecuatoriana de Indígenas Evangélicos (FEINE); Federación Nacional de Organizaciones Campesinas, Indígenas y Negras (FENOCIN); Federación Indígena y Campesino de Imbabura (FICI); Federación de Organizaciones Indígenas de la Faldas del Chimborazo (FOCIFC); Instituto Científico de Culturas Indígenas (ICCI); Movimiento Indígena y Campesino de Cotopaxi (MICC); Pueblo Kayambi; Organización de Pueblos Indígenas de Pastaza (OPIP); Unión de Organizaciones Campesinas Indígenas de Cotacachi (UNORCAC).*

The Ache Heritage Foundation Akha people of Northern Thailand http://www.akha.org/

Alliance for Taiwan Aborigines www.taiwandocuments.org/ata.htm

Alianza Mundial de los Pueblos Indígenas-tribales de los Bosques Tropicales (International Alliance of the Indigenous-Tribal Peoples of the Tropical Forests) http://www.international-alliance.org

Amazon Alliance http://www.amazonalliance.org/index.html

Amazon Watch http://www.amazonwatch.org/.

Amerindian Peoples Association of Guyana (APA) http://www.sdnp.org.gy/apa/

Amnesty International www.amnesty.org/

Ashaninka Peoples of the Peruvian Amazon Website http://www.rcp.net.pe/ashaninka/

Assembly of First Nations (Canada) http://www.afn.ca/

Association on American Indian Affairs www.indian-affairs.org/

Aztec Network http://www.azteca.net/aztec/

Aztec Chat Room http://www.azteca.net/aztec/chat/msgchat.shtml

Center for the Support of Native Lands www.nativelands.org/

Confederation of Indigenous Nationalities of Ecuador http://conaie.nativeweb.org/brochure.html

Convention on Biological Diversity, Indigenous Peoples' Organization of the Amazon Basin www.biodiv.org/default.aspx

Cordillera Peoples' Alliance www.cpaphils.org/

Cultural Survival http://www.culturalsurvival.org.

Cultures on the Edge http://www.culturesontheedge.com/index1.html

Development Gateway. online magazine of *Cultures on the Edge* http://www.developmentgateway.org/

Ditshwanelo-Botswana Center for Human Rights Private Bag 00416, Gaborone, Botswana

DoCip - Centre for Documentation, Research and Information www.docip.org/docipInk.html

Ecuarunari Ecuador Runacunapac Riccharimui (Confederación de los Pueblos de Nacionalidad Kichua del Ecuador) Spanish language site. http://ecuarunari.nativeweb.org/es/nosotros/index.html

Endangered Peoples Project www.aanet.ort/committes/cthr/orgindig.htm

Ethnic Minorities Working Group in Vietnam http://www.emwg.org.vn/

First Nations Development Institute www.firstnations.org/

The First Nations Environmental Network (FNEN) (Canada) http://www.fnen.org/

First People of the Kalahari www.iwant2gohome.org/

First Peoples http://www.firstpeoples.org/who/title_page.htm

Forest Peoples Programme (FPP) (*World Rainforest Movement*) http://www.fpcn-global.org/

Garifuna Network (Caribbean) http://www.garinet.com/

Gwich'in Social and Cultural Institute www.gwichin.ca/

Hach Winik Home Page (Lacandon Maya communities of La Selva Lacandona, Chiapas, Mexico) http://www.geocities.com/RainForest/3134/

Human Rights Internet www.hri.ca/

Human Rights Watch www.hrw.org/

Indian Law Resource Center (USA) www.indianlaw.org/

The Indigenous Environmental Network http://www.ienearth.org/

Indigenous Issues Network http://remote4.acdi-cida.gc.ca/innet

The Indigenous Peoples Biodiversity Information Network http://www.ibin.org/

The Indigenous Peoples' Biodiversity Network http://www.ecouncil.ac.cr/rio/focus/report/english/ipbn.htm

The Indigenous Peoples of Africa Co-ordinating Committee http://www.ipacc.org.za/

Indigenous Survival International www.sicc.sk.ca/

International Alliance of Indigenous and Tribal Peoples of the Tropical Forests http://www.international-alliance.org/english/eng_about.htm

Institut international des Droits de l'Homme (International Institute of Human Rights) www.iidh.org/

International Work Group for Indigenous Affairs http://www.iwgia.org/sw617.asp

Inter Press Service News Agency ("Journalism and Communication for Global Change") http://www.ips.org/institutional/

Inuit Circumpolar Council (ICC) (Formerly *Inuit Circumpolar Conference*) http://www.inuitcircumpolar.com/index.php?ID=1&Lang=En ; *ICC Canada* - www.inuitcircumpolar.com;

ICC Greenland - www.inuit.org; *ICC Alaska* - www.iccalaska.org; *ICC Russia* - www.icc.hotbox.ru

Kalahari Peoples Fund www.kalaharipeoples.org/

Kalahari Support Group www.ksg-san.nl/

Kuru Development Trust www.san.org.za/kuru.home.htm

Mapuche People of Chile Spanish language site. http://www.geocities.com/CapitolHill/Senate/7718/

Minority Rights Group International http://www.minorityrights.org/

Naga Peoples Movement for Human Rights www.cwis.org/fwdp/Eurasia/naga.txt (document)

National Indian Youth Council www.niyc-alb.org/

Native American Journalists Association http://www.medill.nwu.edu/naja

Native American Rights Fund www.narf.org/

Native Network (Bolivia) http://www.nativenetworks.si.edu/eng/rose/cefrec.htm

North Australia Research Unit Australia National University, PO Box 41321, Casuarina NT 0811 Australia

The Partnership for Indigenous Peoples' Environment http://www.pipeorg.com/

Physicians for Human Rights www.physiciansforhumanrights.org/

Plataforma de Información del Pueblo Indio Spanish language site. http://www.puebloindio.org/ceacisa.htm

Rainforest Action Network www.ran.org/

Rainforest web http://www.rainforestweb.org/Rainforest_Information/Indigenous_Peoples/South_America/

Sahabat Alam Malaysia (Friends of the Earth supported) www.foe-malaysia.org/

South and Meso American Indian Rights Center http://saiic.nativeweb.org/

Sarawak Peoples Campaign www.rimba.com/spchomepage.html

Sarayacu Community Spanish language site. http://www.sarayacu.com/

Survival International http://www.survival-international.org/index.php

Taiwan Aboriginals Page www.taiwanfirstnations.org/

Tebtebba (Indigenous Peoples' International Centre for Policy Research and Education) http://www.tebtebba.org/

Tribal Circle http://o.webring.com/webring?ring=americanindians;list

The United Confederation of Taino People http://www.uctp.org/English-Ver.htm

United Nations Permanent Forum on Indigenous Issues www.un.org/esa/socdev/unpfii/

Unrepresented Nations and Peoples Organization http://www.unpo.org/

Wayuu website (Venezuela) Spanish language site. http://www.wayuu.pueblosindigenas.org.ve/

Website and CD-ROM, "The Revolution Will Be Digitized."

Zapatista Net of Autonomy and Liberation http://www.actlab.utexas.edu/~zapatistas/zapatista.htmlPARTICIPATE

Websites with General or Wide-ranging Information

Aboriginal Links International http://www.bloorstreet.com/300block/aborintl.htm#3

AFECT (Akha Association for Education and Culture in Thailand) http://www.akhathai.org/; http://www.hani-akha.org/

The Arctic Council http://www.arcticpeoples.org/

Asian People's Directory http://www.kotan.org/asia/directory/index.html

Canada and the Circumpolar World http://www.dfait-maeci.gc.ca/circumpolar/

Caribbean Amerindian Centrelink http://www.centrelink.org/Venezuela.html

The Centre for Indigenous Environmental Resources (Canada) http://www.cier.ca/

The Hilltribe People of Thailand http://www.tayara.com/club/hilltribe.htm

HmongNet http://www.hmongnet.org/

Indian Circle (USA) (Web ring for federally recognized Native American tribes) http://www.indiancircle.com/

Indigenous Knowledge and Peoples http://www.ikap-mmsea.org/ (2006)

Indigenous Peoples' Center for Documentation, Research and Information http://www.docip.org/dociplnk.html

Indigenous Peoples' Secretariat of the Arctic Council http://www.arcticpeoples.org/

The Indigenous Women's Network (Americas and Pacific Basin) http://www.indigenouswomen.org/

INFOKOORI database (Australia) http://awairs.slnsw.gov.au/koori/

Land Rights (Australia) http://www.faira.org.au/lrqarc.html

Latin American Network Information Center http://lanic.utexas.edu/la/region/indigenous/

MEDIACAN (Aboriginal Peoples and Media, Canada) http://klaatu.pc.athabascau.ca/cgi-bin/b7/main.pl?rid=6758

The Mon Information Home Page (Burma and Thailand) .http://cscmosaic.albany.edu/~gb661/

Native American Communication Resources on the Internet http://hanksville.phast.umass.edu/misc/NAmedia.html

Native American Resources http://www.nativeculture.com/lisamitten/indians.html

Native American Studies Association http://facstaff.uww.edu/mohanp/nasa-currentevents.html

Native American Tribal Government Sites: Web Ring http://www.geocities.com/Athens/Styx/6031/tribeing.htm

Native Web http://www.nativeweb.org/info/

NativeWeb Resources - Radio and Television http://www.nativeweb.org/resources/tv_radio/

The Peoples of the World Foundation http://www.peoplesoftheworld.org/index.html

Researching Indigenous Peoples Rights under International Law http://intelligent-internet.info/law/ipr2.html.

UNESCO Web Site Addressing the Status of the Sami Languages http://amacrine.berkeley.edu/finnugr/uralic-table.html

United South and Eastern Tribes http://www.usetinc.org/defaultpage.cfm?ID=6

University of Southern Queensland (Australia) *Kumbari/Ngurpai Lag Higher Education Centre, Online Resources: Indigenous Media.* http://www.usq.edu.au/kumbari/internetres/media.htm

Village of First Nations Forum for Indigenous people http://www.firstnations.com/

WSIS Indigenous Thematic Planning Conference for Tunisia http://www.itu.int/wsis/docs2/thematic/canada/final-report-indigenous.doc

Chat Rooms and Blogs

International/General

Crisscrossed (Blog that "aims to explore and develop social changes through communication," by Christian Kreutz, a "political scientist and knowledge activist" who has worked for the German Parliament.) http://www.crisscrossed.net/

Global Voices Online (Non-profit global citizens' media project, Harvard Law School Berkman Center for Internet and Society) http://www.globalvoicesonline.org/

North America

American Indian Tribe Discussion Forum http://www.americanindiantribe.com/Discuss/Default.htm:
APTN Forums (Aboriginal Peoples Television Network (Canada) http://www.aptn.ca/forums/index.php
American Indian Chat Rooms http://www.december.com/cmc/mag/1998/jul/baird.html

Sámi: Finland, Norway, Sweden and Russia

Aanta Forsgren's Web Site http://www.itv.se/boreale/samieng.htm
Ajjte Website http://jokkmokk.se/ajjte/index.htm
Forum of Sámi Women http://www.same.net/~sami.nissonforum/
Saami Research and Project Database RÁDJU www.arcticcentre.org/radju
Samenet (Sámi language network) http://forum.same.net
Sametingets Bibliotek www.sb.sametinget.se/svenska/indexsv.htm
Sametingets Hemsida (Sámi Parliament site) www.sametinget.se/
Sami language news http://www.nrk.no/nettradio/
Sami Network Connectivity Project http://www.snc.sapmi.net/; http://www.snc.sapmi.net/Project-docs/Saami-Network-Connect
Sami Parliament in Sweden Web Site http://www.sametinget.se/english/index.html
Samitour www.samitour.no/english/3-attraksjoner.html
Sámi University College http://www.samiskhs.no/)
Sami Youth in Finland http://www.same.net/~ssn/en1.html

Indigenous Print Media

North America

Aboriginal Directory of Ontario http://www.aboriginal.com/
Aboriginal Times http://www.aboriginaltimes.com/
Akwesasne Notes http://www.slic.com/~mohawk/notes.htm
Alberta Native News http://www.albertanativenews.com/
Alberta Sweetgrass http://www.ammsa.com/sweetgrass/

Anishinabek News http://www.anishinabek.ca/CU/News/current.asp
American Indian Art Magazine www.aiamagazine.com/
American Indian Culture and Research Journal www.books.aisc.ucla.edu/aicrj.html
American Indian Quarterly http://www.uoknor.edu/aiq/index.html
Beaver Tail Journal http://eot.com/~slwhite/btj.html
Char-Koosta News (Flathead Indian Nation) http://www.ronan.net/~ckn/
The Cherokee Observer (independent) www.cherokeeobserver.org/
The Cherokee One Feather (Eastern Band of Cherokee) www.cherokeeonefeather.ypgs.net/
The Cherokee Messenger (Cherokee Cultural Society of Houston, Texas) http://www.powersource.com/cherokee/
The Chickaloon News Email: www.chickaloon.org/Media/Chickaloon-News.html
The Eastern Door (Newspaper serving the Mohawks of Kahnawake, Canada) http://www.easterndoor.com/
Eastern Woodland Publishing http://www.cmmns.com/mmnn/ewp.html
First Nations Drum - News from Canada's Native Communities http://www.firstnationsdrum.com/
The First Perspective http://www.firstperspective.ca/ http://www.mbnet.mb.ca/firstper/
Grassroots News http://www.grassrootsnews.mb.ca/
Ha-shilth-sa Newspaper (Nuu chah nulth First Nation, BC, Canada) http://www.nuuchahnulth.org/tribal-council/hashilth.html
Indian Country Today www.indiancountrytoday.com/
International Native News http://ultimate.org/sites/miaorg/inn.htm
Kahtou News Online - The Voice of BC's First Nations http://www.kahtou.com/
The Lakota Times www.lakotacountrytimes.com/
Manitoulin Expositor http://www.manitoulin.ca/
Metis Observer Journal online http://michif.dev.kcdc.ca/
Mohawk Drum http://206.25.233.25/~mohawk/home.html
Native American News http://www.bpinews.com
Native Americas Magazine http://nativeamericas.aip.cornell.edu/
Native Journal Online http://www.nativejournal.ca/pages/frameset.html
Native Peoples Magazine http://hogan1.atiin.com/native_peoples/
Native Voice http://www.native-voice.com/

Navajo-Hopi Observer www.navajohopiobserver.com/
Navajo Times www.navajotimes.com/
News from Indian Country www.indiancountrynews.com/
Nuggu Am Newspaper (Quinault Indian Nation, Washington, USA) www.quinaultindiannation.com/nugguam.htm
Nunatsiaq News (Nunavut, Canada) http://www.nunatsiaqnews.com Bilingual: English and Inuktitut, with some French and Inuvialtun.
Ontario Birchbark The Aboriginal newspaper of Ontario (Canada) http://www.ammsa.com/birchbark/index.htm
Raven's Eye The Aboriginal Newspaper of British Columbia and Yukon (Canada) http://www.ammsa.com/raven/
Red Ink http://w3.arizona.edu/~aisp/redink.htm
Redwire Native Youth Magazine www.redwiremag.com
Saskatchewan Sage - The Aboriginal Newspaper of Saskatchewan http://www.ammsa.com/sage/
Say Magazine (Aboriginal youth) www.SAYmag.com
Secwepemc News - Shuswap Nation (Canada) http://www.secwepemc.org/secnews.html
Sicangu Sun-Times (Rosebud, South Dakota, USA) www.sicangusun-times.com/
Sho-Ban News (Idaho) www.shobannews.com
The Spike (Eastern PowWow information) http://www.thespike.com
Tekawennake Six Nations, Ontario (Canada) http://www.tekanews.com/
War Drum Studios (Native American Comic Books) www.wardrumstudios.com/
Wawatay News http://www.wawatay.on.ca
Whispering Wind Magazine http://www.whisperingwind.com
Wikwemikong Newspaper (The 'Wiky' News) (Wikwemikong First Nation, Canada) http://www.wiky.net/wikydevelopment/Communications/Wiky%20news/wiky_news.htm
Windspeaker (National Aboriginal News, Canada) http://www.ammsa.com/WINDHOME.html
Wolf Howls Journal (Canada) http://tng.fngov.bc.ca/wolfhowls
Wulust Grand Council Newsletter (Tobique First Nation, Canada) http://personal.nbnet.nb.ca/pesun/

Sámi - Finland, Norway, Sweden, and Russia

Assu (newspaper) www.finnmark.net/assu/
Min Áigi (newspaper) www.minaigi.no
Sagat (newspaper) http://www.sagat.no/
Samefolket (magazine) www.info.samefolket.se/index.php
Sámi Press www.samipress.net

Australia

Goolarri Media Enterprise Journal of Australian Indigenous Issues http://www.gme.com.au/
Koori Mail (newspaper) http://www.koorimail.com/
National Indigenous Times http://www.nit.com.au/

Indigenous Media Organizations and Networks: Radio, Television, Film, New Media, Multimedia, and Digital Resources

International/Multi-regional

AMARC, World Association of Community Radio Broadcasters English, French, and Spanish www.amarc.org
Circle of Stories http://www.pbs.org/circleofstories/
Native Indian/Inuit Photographers' Association http://creative-spirit.com/
Native Newmedia Site www.cyberpowwow.net
Native Solidarity News (Radio) http://nsn.tao.ca/home.html
Somena Media: An Indie Aboriginal News and Media Outlet. http://somenamedia.blogspot.com/2005/12/reporting-on-indian-affairs-is-it.html
Turtle Island Native Network http://www.turtleisland.org
UrbaNative Radio for the Urban Native http://www.urbanative.chagoosh.com/
Yahoo! News Full Coverage - First Nations News (Canada) http://ca.fullcoverage.yahoo.com/fc/Canada/First_Nations/

North America

Aboriginal Multi-Media Society http://www.ammsa.com/dsp_login.asp
Aboriginal peoples Television Network http://www.aptn.ca/ National broadcaster (Canada)
Northern Native Broadcast Access Program http://www.canadianheritage.gc.ca/progs/pa-app/progs/paanr-nnbap/broadcast/7_e.cfm
Aboriginal Voices Radio www.aboriginalvoices.com
AIROS Native American Public Telecommunications http://www.airos.org/stations/native.html

AIROS affiliates and Other Native American Radio Stations:

Alaska

KBRW Whalers Radio 680 AM (Barrow) http://www.kbrw.org/
KC 88.1 FM Kashunaniut School District
KIYU 910 AM (Galena) Community Radio Serving the Interior of Alaska - http://www.kiyu.com/
KNBA Koahnic Broadcast Corporation 90.3 FM (Anchorage) http://www.knba.org/
KOTZ (Kotzebue) 720 AM - http://www.kotz.org/
KSDP 830 AM (Sand Point) http://www.ksdpradio.org/
KSKO 870 AM (McGrath)
KUHB 91.9 FM (St. Paul)
KY 640 AM (Bethel)
KZPA 900 AM (Gwandak Public Broadcasting Inc.)(Fort Yon)

Arizona

KGHR 91.5 FM (Tuba City)
KNNB 88.1, 89.9, 99.1 FM (Apache Radio Broadcasting Corp. Whiteriver)
KOHN 91.9 FM (Sells)
KPYT-LP 100.3 FM (Tuscon)
KUYI 88.1 FM (Hotevilla)

California

KIDE 91.3 FM (Hoopa)

Colorado

KRZA 88.7 FM (Alamosa)
KSUT 91.3 FM (Ignacio)

Montana

KGVA 88.1 FM (Harlem)

New Mexico

*KABR 1500 AM (*Magdalena)
KCIE 90.5 FM (Dulce)
KSHI 90.9 FM – (Zuni)
KTDB 89.7 FM (Pinehill)

North Dakota

KABU 90.7 FM [Fort Totten)
KEYA 88.5 FM (Belcourt)
KMHA 91.3 FM (New Town)

Oregon

KWSO 91.9 FM (Warm Springs)

South Dakota

KILI 90.1 FM (Porcupine)
KLND 89.5 FM (McLaughlin)

Washington

KYNR 1490 AM (Confederated Tribes and Bands of the Yakama Nation, Toppenish, Washington)

Wisconsin

WOJB 88.9 FM (Hayward)

Wyoming

KWRR 89.5 FM (Kinnear)

Association of Native Development in the Performing Visual Arts www.andpva.com

Beesum Communications Inc. http://www.beesum-communications.com/index.html

British Columbia Metis Communications Centre http://www.angelfire.com/bc2/bcmetisforum/

Broadband for Rural and Northern Development Pilot Program (Canada) http://www.broadband.gc.ca.

Canadian Aboriginal Media Co-operative (CAMCO) http://www.camco-op.com/

CFWE 89.9 FM (The Aboriginal Multi-Media Society) www.ammsa.com/cfwe/

CHON FM radio and NEDAA television, Northern Native Broadcasting, Yon (NNBY) http://www.nnby.net/main.php

Connecting Canadians - Canadian Content On-line: Aboriginal Digital Collections (ADC) http://www.connect.gc.ca/en/420-e.shtml

Dibaudjimoh - Bringing the News of the Chippewas of Nawash on the Web http://www.bmts.com/~dibaudjimoh/

The Drum (Manitoba and Northwest Ontario) http://www.manitoba-drum.com/

Eagle Feather News http://www.eaglefeathernews.com/

En'owkin Centre www.enowkincentre.ca

First Nations Broadcasting. Native American music and news from Alaska and northern Canada http://www.palmsradio.com/upnorth.html
Includes audio links to:
KNBA Anchorage, Alaska
CBC Inuvik, Canada
KINY 800 AM, Juneau, Alaska
KIAM, Nenana, Alaska (Christian programming)
CFWH 570 AM, Whitehorse, Yukon, Canada
NNN, National Native News

First Nations Confederacy of Cultural Education Centres http://www.fnccec.com/

First Peoples Cultural Foundation www.fpcf.bc.ca

Gryphon Productions Ltd. http://www.gryphonproductions.com/

IMAGe Nation Indigenous Media Arts Group http://www.imag-nation.com/

ImagineNative Media Arts Festival (Centre for Aboriginal Media) www.aboriginalmedia.org

Independent Film and Video Alliance (Canada) www.culturenet.ca/ifva

Indigenous Arts Service Organization www.geocities.com/iaso_1/7.html

Inuit Broadcasting Corporation http://www.inuitbroadcasting.ca/english/index.html

Inuit and Cree Radio and TV Community Broadcasters http://members.tripod.com/~media002/nunamedia.htmlh

Keyah Productions (NNBY) http://www.nnby.net/keyah.htm

Kuh-ke-nah Network of SMART First Nations (K-Net) http://smart.knet.on.ca/

KWE Radio 105.9 FM Mohawk Nation Radio, Tyendinaga http://www.tyendinaga.net/kweradio/kwelive.htm

Micmac-Maliseet Nations News Association http://www.cmmns.com/mmnn/index.html

Missinipi Broadcasting Corporation (MBC) - Saskatchewan Aboriginal Communications http://www.mbcradio.com/index1.html

National Aboriginal Communications Society http://www.ammsa.com/ams/amscanadapubs.html

Native Nations Network (NNN) www.nativenationsnet.net or www.turtle-island.com

Navajo Nation http://www.navajo.org/index.htm

NCI FM: The Spirit of Manitoba Native Communications, Inc. http://www.ncifm.com/

Northern Native Broadcasting, Yukon (NNBY) http://www.nnby.net/

Nunavik (Northern Quebec) community radio stations:

- *Ivujivik Community Radio Appalimmiut Tusautinga FM* (Ivujivik)
- *Kuujjuarapik FM Cree Community Radio* (Kuujjuarapik)
- *Akulivik FM* (Akulivik)
- *Kuujjuaq FM Community Radio Station* (Kuujjuaq)
- *Aupal FM Community Radio Station* (Aupal)
- *Quaqtaq FM Community Radio Station-* (Quaqtaq)
- *Kangirsuk FM Community Radio Station* (Kangirsuk)
- *Inukjuak FM Community Radio Station* (Inukjuak)
- *Salluit Municipal FM Community Radio Station* - Salluit, Quebec, J0M 1S0
- *Kangiqsujuaq - Qakkalik Community FM Station* (Kangiqsujuaq)

Tasiujaq Community Radio FM Station-(Tasiujaq)
Kuujjuarapik Community Radio Station (Kuujjuarapik)
Umiujaq FM Community Radio Station (Umiujaq)

Nunavut Environmental Database, on-line http://ned.nunavut.ca

Omaha Indian Music http://memory.loc.gov/ammem/omhhtml/

Shenandoah Films Distributor http://www.shenandoahfilms.com

Siku Circumpolar News Service http://www.sikunews.com/

Sleeping Buffalo Internet Radio - 88.8 FM http://www.banffcentre.ab.ca/Aboriginal_Arts/sleepingbuffalo/

Smoke Signals First Nations Radio, CHRW 94.9 FM London, Ontario, Canada; www.chrwradio.com; http://wwwachannel.ca/london/promo/SmokeSignals/index.html, http://smokesignals.netfirms.com; http://peterdawson.typepad.com/blog/smokessignals/index.html

Society of Communication Atikamekw - Montagnais (SOCAM) http://www.meskino.qc.ca/socam/indexe.html

Taiga Communications, Inc. http://www.taiga-communications.com/main_taiga.htm; http://www.firstperspective.ca

Taqramiut Nipingat incorporated http://www.taqramiut.qc.ca/en/index.htm

Thunder Voice News http://www.abcentre.org/serv18.html

Vancouver's Independent Media Organization vancouver.indymedia.org

WOJB - Ojbway Radio http://www.wojb.org/

Latin America

Centro de Investigaciones y Estudios Superiores en Antropología Social-CIESAS CIESAS (Center of Research and Study of Social Anthropology) Spanish language site www.ciesas.edu.mx/

Centro de Derechos Humanos de la Montaña Tlachinollan (Center of Human Rights from the Mountain Tlachinollan) Spanish language site http://www.tlachinollan.org/

La Casa del Escritor Sna Jtz'ibajom (The House of the Writer Sna Jtz'ibajom) English language site http://pages.prodigy.net/gbonline/casaescr.html

Frente Indigena Oaxaqueño Binacional (Binational Center for the Development of the Indigenous Communities) Bilingual: English and Spanish. http://www.fiob.org/

Fundación Comunalidad A.C. (Comunalidad Foundation) Spanish language site. http://espora.org/biblioweb/Comunalidad/

Grupo de Estudios Sobre la Mujer "Rosario Castellanos" (Group of Studies on Women "Rosario Castellanos") Spanish language site. www.laneta.apc.org/oaxaca/genero/casamuje.html

Museo Nacional de Culturas Populares (National Museum of Popular Cultures) Spanish language site www.cnca.gob.mx/cnca/popul/mncp.htm

Programa La Neta Spanish language site. http://www.laneta.apc.org/

Proyectos mapuches (includes the exhibition, *Fotografia mapuche: Construcción y montaje de un imaginario)* Spanish language site: www.puc.cl/proyectos/mapuches/html/

Unión de Comunidades Indígenas de la Zona Norte del Istmo (Union of Indigenous Communities of the Northern Region of the Isthmus of Tehuantepec Spanish language site. www.mesoamericaresiste.org/

Sistema de Radiodifusoras Culturales Indigenista Indigenous Broadcasting System. Spanish language site. www.cdi.gob.mx/index.php?id_seccion=23

XECTZ 1350 AM Sierra Norte, Mexico (Flagship station of the *Sistema de Radiodifusoras Culturales Indigenista*) Broadcasts in Indigenous languages to more than 1,000 locations in the mountain range between the states of Puebla and Veracruz) http://civila.com/mexico/inipue2/Index.html

Australia

ABC Indigenous Program Unit http://abc.net.au/message/

Australian Film Commission Indigenous films http://www.afc.gov.au/industrylinks/screenorg/indigenous.aspx

Awaye (ABC Radio) http://www.abc.net.au/message/radio/awaye/

Central Australian Aboriginal Media Association http://www.caama.com.au/caama/

Imparja Television http://www.imparja.com.au/

Message Stick ABC Television online Indigenous newsletter http://www.abc.net.au/message/

National Indigenous Media Association of Australia http://www.nimaa.org.au/

Pitjantjatjara Yankunytjatjara Media (PY Media) http://waru.org/organizations/pymedia/
Speaking Out ABC Radio http://www.abc.net.au/message/radio/speaking/
Top End Aboriginal Bush Broadcasting Association (TEABBA) http://www.teabba.com.au/
Waringarri Media 6WR http://www.indiginet.com.au/
Warlpiri Media Association http://www.warlpiri.com.au/

Māori: Aotearoa/New Zealand

Aotearoa Cafe Forums Māori Chat Rooms http://www.aocafe.com
Aotearoa Maori Internet Organization http://www.amio.maori.nz
From Hawaiiki to Hawaiiki (The first Maori website, founded in 1995) http://www.maaori.com
irirangi.net (Portal to Te Reo Maori radio on the Internet) http://www.irirangi.net
Maori.gen.nz (Group of Māori websites and links to portals) http://www.maori.gen.nz/
Maori News Online and Maori Portal - Te Karere Ipurangi http://maorinews.com/karere/
Maori Television http://www.maoritelevision.com/
Te Kāea Email: tekaea@maoritelevision.com
Ruia Mai: Maori Radio Māori language broadcaster http://www.ruiamai.co.nz/
Te Karere Ipurangi http://www.maorinews.com/karere/
Television New Zealand (TVNZ) (NGa Matatiki Rorohiko: Maori Electronic Resources, Television and radio) www.tvnz.co.nz/
Te Putatara (Māori) http://www.putatara.net
Tuhono (Network of individuals) http://www.tuhono.net
Waatea603am Urban Maori Radio Bilingual broadcasts. http://www.waatea603am.co.nz/

Ainu: Japan

Ainu Association http://www.ainu-assn.or.jp/index.html
Ainu dance and music http://www.sh.rim.or.jp/~moshiri/

Ainu: Spirit of a Northern People Smithsonian Institution site (USA/Japan) http://www.s2nmedia.com/news/; http://www.mnh.si.edu/arctic/ainu/index.html, http://www.ethnologue.com/show_language.asp?code=ain, http://www.frpac.or.jp/eng/e_itm/project04_1.html

Foundation for Research and Promotion of Ainu Culture (FRPAC) http://www.frpac.or.jp/eng/index.html; http://www.frpac.or.jp/eng/e_itm/project03.html

Hokkaido Ainu Cultural Research Centre http://www.pref.hokkaido.lg.jp/ks/abc/index.htm

Industrial Arts Association http://www.marimo.or.jp/~akanainu/

Radio Sapporo Ainu Language Lessons http://www.stv.ne.jp/radio/ainugo/index.html

Greenland

Culture Greenland website (Sisimiut Museum) Greenlandic language version: http://www.kulturi.gl

Danish version: http://www.kultur.gl English version: http://www.culture.gl

National Picture Database of Greenland. http://www.culture.gl/picture

Sermitsiaq (Independent newspaper, founded in 1958) Published in Greenlandic, Danish, and English; English language link: http://sermitsiaq.gl/english/

Kalaallit Nunaata Radio (KNR) Radio and Television Greenland http://www.randburg.com/gr/radiotvg.html

Sámi: Finland, Norway, Sweden, Russia

Norwegian Broadcasting Corporation (NRK) English Language site http://www.nrk.no/about/

NRK Sámi Radio Sámi, Norwegian, Finnish, German, English and Spanish www.nrk.no/samiradio/

Sámi Radio online newspaper www.samiradio.org

YLE Sámi Radio, Finland English language site http://lotta.yle.fi/srwebanar.nsf/sivut/frontpage; www.yle.fi/samiradio/

Social Networking Websites

An increasing number of Indigenous people, especially adolescents and young adults, are creating their own contact pages and networks on an ever-expanding array of social networking websites. Users can create personalized sites with music, videos, photographs, opinions, news, and blogs, and can communicate with other users over the Internet. The following are among the most prominent sites:

Facebook www.facebook.com/, A general service that allows the user to create a personal contact page, and to communicate with "friends," over the Internet, and to expand the individual's personal network by linking friends with each other.

Flickr www.flickr.com/, A service designed primarily for the exchange of photographs and other images over the Internet.

MySpace www.myspace.com/, Widely used by artists and musicians to share and showcase their work, and increase their Industry contacts.

Twitter www.twitter.com/, A global chat room that allows individuals to "tweet" in real-time with innumerable known and unknown others.

Notes

Notes to Preface

1. Browne's work is important for its consideration of media and peoples in several parts of the world, government policies, and important outcomes, such as helping to revive Indigenous languages. He visited media organizations and some of the prime movers (e.g., in Sámi and Mohawk radio). He misses the centrality of Canada in the global picture, having focused on Mohawk communities and briefly noted policy developments, but not the scope and widening influence of Indigenous northern broadcasting, particularly TVNC (Browne 1996: 44). Most surprisingly, he refers to first peoples in Canada as "Native Americans" (Browne 1996: 5), implying that Canada is an extension of the United States and leaving First Nations people, Métis, and Inuit out in the cold.

Notes to Acknowledgements

1. His surname has been spelled in two ways, as Whiteye and White Eye.

Notes to Chapter 2

1. The origins of wireless radio transmission remain contentious. One version is that wireless radio transmission began with a broadcast of hymns sung by the Canadian inventor, Reginald Fessenden, on Christmas Eve, 1906, sent by radio from Massachusetts to Scotland (CBC Digital Archives 1979). However, Fessenden's first broadcast had been earlier, on 23 December 1900, when he sent a wireless telegraph message between two towers near Washington, DC. Fessenden is often upstaged by the better-known Ital-

ian inventor, Guglielmo Marconi, who "claimed to have sent an even earlier wireless telegraph in 1895." In 1979, the Canadian Broadcasting Corporation televised a debate about the claims and history of the two inventors (CBC Digital Archives 1979).

2. This piece was distributed and discussed at the Fourth Cinema Workshop, Leeds School of English, Humanities Research Institute, University of Leeds, England, at which I was privileged to speak, along with the featured filmmakers, Barry Barclay (Māori, Aotearoa/New Zealand) and Jeff Bear (Maliseet, Canada), 24 October 2005.

Notes to Filmography

1. My knowledge of the "case" depicted in "The Story of Joe and Elise" is personal as well as professional. Elise was originally to have co-authored the study of Inuit naming that I ended up completing after her death (published in 2007 as *Names and Nunavut: Culture and Identity in the Canadian Arctic*, New York and Oxford: Berghahn Books, and in a brief, earlier version as *Names, Numbers and Northern Policy*, Halifax: Fernwood). Her loss is still keenly felt in Nunavut, in Pauktuutit, (Inuit Women's Association) and among Inuit across Canada.

Bibliography

ABC. July 2007. "New Indigenous TV station turns on." Australian Broadcasting Corporation, July 13.

———. November 2007. "Rudd 'backtracked on Indigenous pledge.'" ABC, November 12.

———. 2005. *Message Stick Aboriginal News Online*, June 17.

———. 2005. "Voices of Australia." *Message Stick Aboriginal News Online*, October 31.

———. 2005. "Egan backs Aboriginal out-stations." *Message Stick Aboriginal News Online*, December 16.

Aboriginal Tourism Australia. 2004. *Welcome to country: Respecting Indigenous culture for travelers in Australia.* Australian Government, Department of the Environment and Heritage.

Aboriginal Voices. 1999. Cover photograph and caption. October/November.

AboriNews.com. 2005. Guatemala Indigenous Community Radio Project." Cultural Survival, October 3.

Africa-in-motion. 2009. (online) http://www.africa-in-motion.org.uk/

The Age online, www.theage.com.au/. November 28.

Aikio-Puoskari, U. 1998. "Sámi Language in Finnish Schools." In *Bicultural Education in the North*, ed. E. Kasten, 47–57. New York/Munchen/Berlin: Waxmann Münster.

Alfred, T. G. 1999. *Peace Power Righteousness: an indigenous manifesto.* Don Mills, Ontario: Oxford University Press.

Alia, V. 2007. *Names and Nunavut: Culture and Identity in the Inuit Homeland.* New York and Oxford: Berghahn Books.

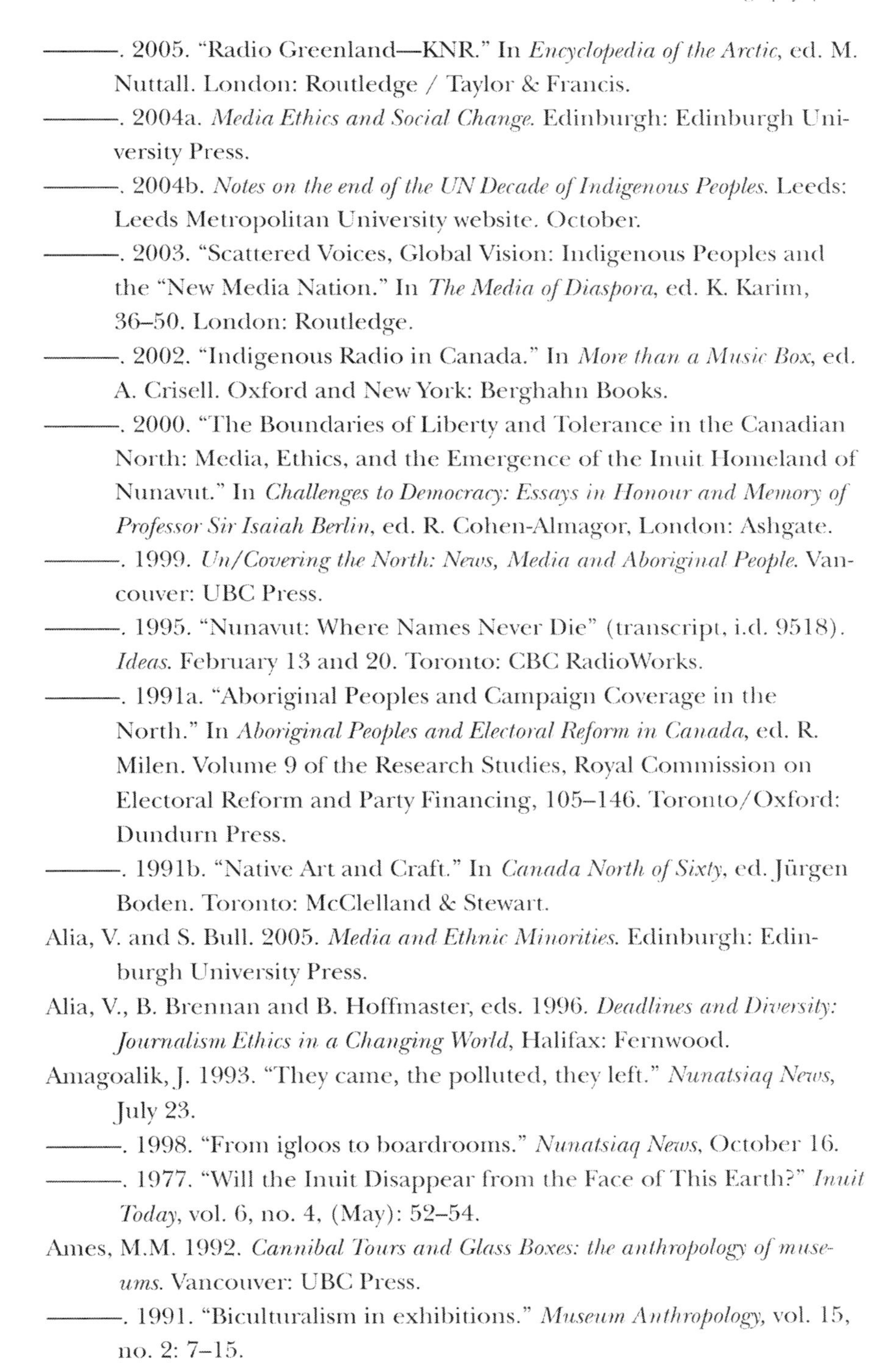

———. 2005. "Radio Greenland—KNR." In *Encyclopedia of the Arctic*, ed. M. Nuttall. London: Routledge / Taylor & Francis.

———. 2004a. *Media Ethics and Social Change.* Edinburgh: Edinburgh University Press.

———. 2004b. *Notes on the end of the UN Decade of Indigenous Peoples.* Leeds: Leeds Metropolitan University website. October.

———. 2003. "Scattered Voices, Global Vision: Indigenous Peoples and the "New Media Nation." In *The Media of Diaspora*, ed. K. Karim, 36–50. London: Routledge.

———. 2002. "Indigenous Radio in Canada." In *More than a Music Box*, ed. A. Crisell. Oxford and New York: Berghahn Books.

———. 2000. "The Boundaries of Liberty and Tolerance in the Canadian North: Media, Ethics, and the Emergence of the Inuit Homeland of Nunavut." In *Challenges to Democracy: Essays in Honour and Memory of Professor Sir Isaiah Berlin*, ed. R. Cohen-Almagor, London: Ashgate.

———. 1999. *Un/Covering the North: News, Media and Aboriginal People.* Vancouver: UBC Press.

———. 1995. "Nunavut: Where Names Never Die" (transcript, i.d. 9518). *Ideas.* February 13 and 20. Toronto: CBC RadioWorks.

———. 1991a. "Aboriginal Peoples and Campaign Coverage in the North." In *Aboriginal Peoples and Electoral Reform in Canada*, ed. R. Milen. Volume 9 of the Research Studies, Royal Commission on Electoral Reform and Party Financing, 105–146. Toronto/Oxford: Dundurn Press.

———. 1991b. "Native Art and Craft." In *Canada North of Sixty*, ed. Jürgen Boden. Toronto: McClelland & Stewart.

Alia, V. and S. Bull. 2005. *Media and Ethnic Minorities.* Edinburgh: Edinburgh University Press.

Alia, V., B. Brennan and B. Hoffmaster, eds. 1996. *Deadlines and Diversity: Journalism Ethics in a Changing World*, Halifax: Fernwood.

Amagoalik, J. 1993. "They came, the polluted, they left." *Nunatsiaq News*, July 23.

———. 1998. "From igloos to boardrooms." *Nunatsiaq News*, October 16.

———. 1977. "Will the Inuit Disappear from the Face of This Earth?" *Inuit Today*, vol. 6, no. 4, (May): 52–54.

Ames, M.M. 1992. *Cannibal Tours and Glass Boxes: the anthropology of museums.* Vancouver: UBC Press.

———. 1991. "Biculturalism in exhibitions." *Museum Anthropology*, vol. 15, no. 2: 7–15.

———. J.D. Harrison, and T. Nicks. 1988. "Proposed museum policies for ethnological collections and the peoples they represent." *Muse*, vol. 6, no. 3: 47–52.

Anderton, A. 2007. "Heart of the Matter." *Cultural Survival Quarterly*, summer, 18–23.

Andrews, M. 2006. "Aboriginal radio station to set up shop." *Vancouver Sun*, May 25.

Ang, I. 1996. *Living Room Wars: Rethinking Media Audiences for a Postmodern World.* London: Routledge.

Appadurai, A. 2000. *Modernity at Large: Cultural Dimensions of Globalization.* Minneapolis: University of Minnesota Press.

———. 1990. "Disjuncture and Difference in the Global Cultural Economy." *Theory, Culture & Society*, vol. 7: 295–310.

———. 1988. "Putting Hierarchy in Its Place." *Cultural Anthropology*, vol. 3, no. 1: 36–49.

APTN. 2008a. *Aboriginal Peoples Television Network* (brochure).

APTN. 2008b. "Aboriginal Music Invasion will repeat the success of British Invasion and Motown." Aboriginal Peoples Television Network website, www.aptn.ca/, August 27.

APTN. 2006. "Viewing audiences continue to grow." Aboriginal People's Television Network website, January 31.

APTN. 2005. "APTN Bingo & a Movie" (promotional flyer and bingo cards), Aboriginal Peoples Television Network, January–February.

———. 2006. "APTN Directors elect new chair and new members." APTN Press release, APTN website, www.aptn.ca/, January 26.

———. 2005. "APTN launches the Aboriginal Media Education Fund." APTN Press release, Winnipeg, November 25.

———. 2001. Press release, website, http://www.aptn.ca/en/CRTCJan5-01.htm.

Aqpik, O. 2004. *New Made-in-Nunavut School Books Launched.* October 4. Iqaluit: Department of Education.

Arbelaitz, L. 2007. "The Silent Revolution." Gáldu: Resource Centre for the Rights of Indigenous Peoples website. <http://www.galdu.org/>.

Armitage, Andrew. 1995. *Comparing the Policy of Aboriginal Assimilation: Australia, Canada, and New Zealand.* Vancouver: UBC Press.

Arnait Video Productions Collective. 2008. *New Inuktitut-Language Film:*"Before Tomorrow." Igloolik, Nunavut, September 22. <http://www.cbc.ca/canada/north/story/2008/09/16/arnait-tiff.html>.

Assembly of First Nations. 2009. Website. http://www.afn.ca/.

Austen, I. 2008. "Canada Offers an Apology for Native Students' Abuse." *New York Times*, June 12 (online), www.nytimes.com/.

Backhouse, C. 1999. *Colour-Coded: A Legal History of Racism in Canada, 1900–1950.* University of Toronto.

Balbulican. 2007. "Dr. Gail Guthrie Valaskakis, 1939–2007." *Stageleft: Life on the [lower] left side* (blog site). July 24.

Banning, K. and T. Walcott, eds. 1995. "Mau Mauing Multiculturalism." *Border/Lines*, April, Issue no. 36.

Barclay, B. 2005. *Mana Tuturu: Maori Treasures and Intellectual Property Rights* (excerpts from draft of book, July 6).

———. 1996. "Amongst Landscapes." In *Film in Aotearoa New Zealand*, 2nd ed., ed. J. Dennis and J. Bieringa. Wellington, New Zealand: Victoria University Press.

Bartels, D.A. and A.L. Bartels. 1995. *When the North Was Red: Aboriginal Education in Soviet Siberia.* Canada: McGill-Queen's University Press.

Battarbee, K. 2007. *Speaking Cree on the Cellphone: Heritage Languages in an ICT Environment.* Unpublished paper, British Association of Canadian Studies annual conference, From Blueberries to Blackberries: Traditions and Technologies in Canada, April 11–13, Durham: St. Aidan's College, University of Durham (UK).

Bayagul. 2007. Description of an exhibition in the Powerhouse Museum, Sydney, Australia: *Bayagul Contemporary Indigenous Communication.* Powerhouse Publications website, www.powerhousemuseum.com/.

BBC. 2006. BBC website, www.bbc.co.uk/.

———. 2005. British Broadcasting Corporation (BBC) World News, July 15.

———. 2005. BBC website

———. 1998. "The Jerusalem Report." *BBC News.* British Broadcasting Corporation, August 31.

Beckett, J. 1987. *Torres Strait Islanders: custom and colonialism.* Sydney: Cambridge University Press.

Beckett, J. 1990. "The History of the Torres Strait Islands—Australian Relationship." In *The Strategic Significance of Torres Strait*, ed. Ross Babbage, 240–256. The Australian National University, Canberra: Strategic and Defense Studies Center, Research School of Pacific Studies.

Belich, J. 1996. *Making Peoples: A History of the New Zealanders from Polynesian Settlement to the End of the 19th Century.* Auckland: Penguin.

Bell, D. V. J. 1975. *Power, Influence, and Authority: An Essay in Political Linguistics.* New York: Oxford University Press.

Bell, J. 2004. "Nunavut's population nears 30,000." *Nunatsiaq News* online, June 18.

——— 1993. "The Real World." Editorial. *Nunatsiaq News,* July 16, 2.

Benamrane, D. 2000. Rural Communication: A Strategic Link for Poverty Alleviation in Niger. FAO First International Workshop on Farm Radio Broadcasting. <http://www.fao.org/docrep/003/x67213/x6721321.htm>.

Berryhill, P. 1995. "Afterword: A Native Producer's Perspective." In *Signals in the Air: Native Broadcasting in America,* ed. M.C. Keith, 149–154. Westport, CT: Praeger.

Berthelsen, T.O.K. 1995. "Greenland Home Rule." *Indigenous Affairs* 1: 14–20.

Bernier, H. and P. Barnsley. 2008. Live television coverage of the government apology to Aboriginal people. APTN, June 11.

Bindé, J. 2005. *Towards Knowledge Societies.* UNESCO World Report. Paris: UNESCO.

Bjørklund, I. 2000. *Sápmi—becoming a nation.* Tromsø: Tromsø University Museum.

Bodley, J. 1994. *Victims of Progress.* Mountain View: Mayfield Publishing Company.

Boomfield, S. 2007. "Boom in blogs gives Africans a voice on the Web." *The Independent,* August 2, 26.

Bourdieu, P. 1991. *Language and Symbolic Power.* Cambridge, MA: Harvard University Press.

Brooten, L. 2008. "Media as our Mirror: Indigenous Media of Burma (Myanmar)." In *Global Indigenous Media: Cultures, Poetics, and Politics,* ed. P. Wilson and M. Stewart, Durham, North Carolina: Duke University Press, 111-127.

Brown, D. L. 2008. "Canadian Government Apologizes For Abuse of Indigenous People." *Washington Post,* June 12, A01.

———. 2002. "Trail of Frozen Tears: The Cold War is over, but to Native Greenlanders displaced by it, there is still no peace." *Washington Post.*

Brown, J. 2005. "World Summit on the Information Society." *Cultural Survival Voices,* vol. 4, no. 1, June 15.

Browne, D. R. 1996. *Electronic Media and Indigenous Peoples: A Voice of Our Own?* New York: John Wiley and Sons.

Buddle, K. 2008. "Transistor Resistors: Native Women's Radio in Canada and the Social Organization of Political Space from Below." In *Global Indigenous Media*, 129-144.

Burgess, M. and G. Guthrie Valaskakis. 1992. *Indian Princesses and Cowgirls: stereotypes from the frontier.* Montreal: OBORO.

Calder, P. 1996. "Would-be warriors—New Zealand Film since the Piano" In *Film in Aotearoa New Zealand*, 2nd ed., ed. J. Dennis and J. Bieringa. Wellington: Victoria University Press.

Campbell, R. 1996. "Nine Documentaries." In *Film in Aotearoa New Zealand*, 2nd ed., ed. J. Dennis and J. Bieringa. Wellington: Victoria University Press.

Campion, C. 2007. "Arctic Magic." *The Observer Music Monthly*, no. 41, January: 42–47.

Canada, Department of Indian and Northern Affairs. 1972. *Canada's North: 1970–1980.* Ottawa: Department of Indian and Northern Affairs.

Canada NewsWire. 2001.

Canadian Press. 2007. "UN set to adopt native rights declaration, no thanks to Canada: critics." Canadian Press (accessed through *CBC News online*), www.thecanadianpress.com/. September 6.

CBC (Canadian Broadcasting Corporation). 2007. "First Nations women use 'photo voice' to tell stories." *CBC Arts online*, February 1.

———. 2007. "Cherokees revoke tribal membership of slave descendants." The Canadian Press and the Associated Press, *CBC News online*, March 4.

———. 2007. "Inuit artist accuses CRA staff of writing racist tax memo." *CBC News online*, October 26.

———. 2006. "Photo collection of premiers' homes 'great Canadiana.'" *CBC Arts online*, December 18.

———. 2006. "Haida 'front and centre' at indigenous people's conference." *CBC Arts online*, June 19.

———. 2005. "For first time PM invites First Nations leaders to top meeting." *CBC News Online*, October 21.

———. 2004. "Nunatsiavut: Our beautiful land." *CBC News Online*, July 2.

CBC Digital Archives. 1979. Television broadcast about the origins of wireless radio telegraphy. October 16. <http://archives.cbc.ca/science_technology/technology/clips/10258> (accessed 6 September 2008).

Carroll, R. and A. Schipani. 2009. "Bolivia's 'little Indians' find voice." *The Observer*, 26 April, 34.

Christensen, N. B. 2003. *Inuit in Cyberspace: Embedding Offline Identities Online*. Copenhagen: Museum Tusculanum Press, University of Copenhagen.

CineFocus Canada. 2007. Online (Toronto) TV documentary co.

Cherrington, M. 2007. "The Last Word." *Cultural Survival Quarterly*, 31:2, July: 14.

———. 2006. "Guatemala Radio Project Goes on the Air." *Cultural Survival Quarterly*, spring: 14.

Chikkap, M. 1993. "We may hear the pulsation of the 21st century beating for our future. It calls for "living together". In *The Committee to Organize the Exhibition of Ainu Pictures to Promote Human Rights*, 13–15.

Clifford, J. 1997. *Routes: Travel and Translation in the Late Twentieth Century*. Cambridge, MA and London: Harvard University Press.

Cohen, A. P. 1985. *The Symbolic Construction of Community*. London and New York: Routledge.

The Committee to Organize the Exhibition of Ainu Pictures to Promote Human Rights. 1993. *Exhibition of Ainu Pictures to Promote Human Rights* (catalogue). Sapporo city, Japan: Hatamoto printing Co.

Compact Oxford English Dictionary Online. 2008. "Outlaw." AskOxford.com website. <http://www.askoxford.com/dictionaries/compact_oed/?view=uk> (accessed 5 September 2008).

Cordovoa, A. and G. Zamorano. 2004. Mapping Mexican Media Indigenous and Community Vide*o and Radio—Native Networks*. Washington, D.C.: Smithsonian Institution.

Council on Indigenous Peoples in Taiwan. 2005. Council on Indigenous Peoples in Taiwan, July 15.

Cowan, E. 1969. "Eskimos of Canada: Between Conflicting Cultures." *New York Times*, Sunday, April 13, 22.

Cross, S.C.I. 2006. *Re-imagining Fourth World Self Determination: Indigenous Self-Governance in Greenland, with Implications for Torres Strait*. Ph.D. Thesis, Supervisor: P. Jull, University of Queensland (online), www.uq.edu.au/.

Crow Dog, M. and R. Erdoes. 1992. *Lakota Woman*. New York: Grove Press.

Crowley, D. and P. Heyer. 1991. *Communication in History: Technology, Culture, Society*. White Plains, NY: Longman.

Csonka, Y. and P. Schweitzer. 2004. "Societies and Cultures: Change and Persistence." *Arctic Human Development Report*, 45–68. Akureyri, Iceland: Stefansson Arctic Institute.

CTV. 2006. Press release, April 12. Toronto.

Cullen, T.T. 2005. "Sovereignty Unplugged: Wireless Technology and Self-Governance in the Navajo Nation." *Cultural Survival Quarterly*, vol. 29, no. 2, summer: 32–34.

Cultural Survival. 2008. "Australia Apologizes to Aborigines for Stolen Generations." *Cultural Survival Quarterly*, spring: 7.

——— 2007. "UN Adopts the Declaration on the Rights of Indigenous Peoples." *Cultural Survival Quarterly* 31:3, October: 4–5.

——— 2006. "Promoting Indigenous Participation in Democracy in Guatemala." *Annual Report 2006*, 5. Cultural Survival.

———. 2007. "United States Passes Native American Language Bill." *Cultural Survival Quarterly*, vol. 31, no. 1, spring: 5.

———. 2006. *2006 Annual Appeal leaflet*. Cambridge, MA: Cultural Survival.

———. 2006. *Weekly Indigenous News* (online). February 6.

———. 2006. "Promoting Indigenous Participation in Democracy in Guatemala." *Annual Report 2006*, 5. Cultural Survival.

Cultural Survival e-newsletter. 2009. April.

Cultural Survival e-newsletter. 2008. "Cultural Survival Report Leads to Ainu Recognition." *Cultural Survival e-newsletter*, July–August.

Curry, A. 1998. "Love Lost and Found: Amnesia—the Injured Brain." *Dateline*. New York: NBC News (published online at MSNBC.com).

Curry, B. and G. Galloway. 2008. "We are sorry." *Globe and Mail*, June 12 (online).

Dahl, J. 2001. "Self-Government in Greenland." Indigenous Affairs 3: 36–41.

———. 1986. "Greenland: Political Structure of Self-Government." *Arctic magazine*

Daley, P.J. and B.A. James. 2004. *Cultural Politics and the Mass Media: Alaska Native Voices*. Urbana and Chicago: University of Illinois Press.

Danenhauer, N. Marks and R. Danenhauer. 2004. "Raven Stories: Introduction." In *Voices from Four Directions: Contemporary Translations of the Native Literatures of North America*, ed. B. Swann, 25–41. Lincoln: University of Nebraska Press.

Davidson, M. 2007. "The Lost World." *Disability Now*, February, 24–25.

Davis, R. and M. Zannis. 1973. *The Genocide Machine in Canada: The Pacification of the North.* Montreal: Black Rose.

Delgado-P, G., (2002) "Solidarity in cyberspace: Indigenous peoples online: have new electronic technologies fulfilled the promise they once seemed to hold for Indigenous peoples? The answers are yes, and no" in *NACLA Report on the Americas,* Vol. 35, No. i5 (March), 49-53.

DeMain, P. 2001. "Guerillas in the Media." In *We, the People: of earth and elders,* Volume II, ed. S.L. Chapman, 129–132. Missoula, MT: Mountain Press Publishing Company.

Denmark. 1978. *Greenland Home Rule Act,* No. 577, Section 1. Government of Denmark, November 29.

Development Gateway. 2003. *Mission Empowerment: the Digital Vision Program.* <http://www.developmentgateway.org/hode/133831/sdm/docview?docid=551770>.

Diatchkova, G. 2008. "Indigenous Media as an Important Resource for Russia's Indigenous Peoples." In *Global Indigenous Media,* 214-252.

Dickason, O. P. 1992. *Canada's First Nations: a History of Founding Peoples from Earliest Times.* Toronto: McClelland & Stewart.

Donnan, H. and T.M. Wilson. 1999. *Borders: Frontiers of Identity, Nation and State.* Oxford, NY: Berg.

Dybbroe, S. 1996. "Questions of Identity and Issues of Self-determination." *Etudes/Inuit/Studies,* vol. 20, no. 2: 39–53.

Dyck, N. 1985. *Indigenous People and the Nation State: "Fourth World" Politics in Canada, Australia and Norway.* Newfoundland: Institute of Social and Economic Research Memorial, University of Newfoundland.

Eastern Door. 1999. *Eastern Door* (weekly newspaper), vol. 8, no. 12: April 16. <http://www.easterndoor.com>.

Endeavor Productions. 2006. Endeavor Productions website, www.imdb.com/company/. 2006.

Escobar, A. 1994. "Welcome to Cyberia: Notes on the Anthropology of Cyberculture." In *Current Anthropology* 35(3): 211–232.

Études/Inuit/Studies. 1992. "Inuit Collective Rights and Powers." Special issue, vol. 16, nos. 1–2.

Fanon, F. 1967. B*lack skin, White masks.* New York: Grove Press.

Farmer, G. 1998. Letter from the Editor: Time in a Computer Chip." *Aboriginal Voices,* July/August, 6.

Mutuku, T. 2005. "Fiji Community Radio Initative with Suitcase Radio." WACC (WACC (World Association for Christian Communication)

Pacific Blog, January 17. <http://www.waccglobal.org/de/regions/pacific/pacific_news> (accessed 7 September 2008).

Fienup-Riordan, A. 1995. *Freeze Frame: Alaska Eskimos in the Movies.* Seattle: University of Washington Press.

First Nations Education Council. 2006. Press release, June 21. Wendake, Québec: FNEC.

First Nations Film and Video World Alliance. 1993. *Organizational plan by working group.* Unpublished document distributed internationally at meetings, conferences, and by e-mail and post.

Forsgren, A. 1998. "Use of Internet Communication among the Sami People." *Cultural Survival Quarterly*, winter: 34–36.

Fung, Richard. 1993. "Working Through Cultural Appropriation." *Fuse* 16(5/6), summer: 16–27.

Galtung, J. 1980. *The True Worlds: A Transnational Perspective.* New York: The Free Press.

Gartrell, A. 2007. "Indigenous leaders and activists welcome the end of Howard era and Brough's loss in Longman." Special Election '07 Edition, *National Indigenous Times* (NIT), November 24.

Gaski, H. 1997. "Voice in the Margin: A Suitable Place for a Minority Literature." In *Sámi Culture in a New Era: The Norwegian Sámi Experience*, ed. Harald Gaski, 199–220. Karasjok: Davvi Girgi OS.

Gauthier, J. 2008. "'Lest Others Speak for Us': The Neglected Roots and Uncertain Future of Maori Cinema in New Zealand." In *Global Indigenous Media,* 58-73.

George, J. 2001. "Journalist keeps aboriginal languages alive in Russian Arctic." *Nunatsiaq News*, June 8, 13.

———. 2001. "All Nunavik communities will soon get access to the Internet." *Nunatsiaq News*, March 9, 25, 31.

Ginsburg, Faye. 2008. "Rethinking the Digital Age." In *Global Indigenous Media,* 287-306.

Golubchikova, V.D., Z.I. Khvtisiashvili and Y.R. Akbalyan. 2005. *Practical Dictionary of Siberia and the North* (English ed.). Moscow: European Publications and Severnye Prostory.

Gould, G. 1971. *The Idea of North.* Toronto: Canadian Broadcasting Corporation.

Government of Denmark. 1978. *Greenland Home Rule Act*, No. 577, Section 1, November 29.

Government of the Yukon. 2008. Population Statistics, as of December 2007. Website, www.gov.yk.ca/ (accessed 24 August 2008).

Grace, S. E. 2001. *Canada and the Idea of North.* Montreal and Kingston, Canada: McGill-Queen's University Press.

Grant, B. 1995. *In the Soviet House of Culture: A Century of Perestroikas.* Princeton, NJ: Princeton University.

Greenslade, R. 2007. Report on journalist deaths for BBC *World Agenda magazine*, posted by BBC World Service online, www.bbc.co.uk/worldservice/, December 7.

Greer, S. 2001. "Everett Soop: The pit bull of Native journalism." Association of Canadian Editorial Cartoonists website, www.canadiancartoonists.com/, September.

———. 1999. "I See My Tribe Is Still Behind Me: Everett Soop." *Aboriginal Voices*, Volume 5, October/November, 38–41.

Guly, C. and M. Farley. 2008. "Prime minister apologizes for century of abuses." *The Chicago Tribune.* (Also published as "Canada's Native People Get A Formal Apology," in *The Los Angeles Times*, June 12 (online)).

Gunew, S. 2004. *Haunted Nations: The Colonial Dimensions of Multiculturalisms.* London / New York: Routledge.

Habeck, J. O. 1998. "The Existing and Potential Role of the Internet for Indigenous Communities in the Russian Federation." In *Bicultural Education in the North: Ways of Preserving and Enhancing Indigenous People's Languages and Traditional Knowledge*, ed. E. Kasten, 275–287. Münster: Waxmann.

Haliday, J. and G. Chehak, in cooperation with the Affiliated Tribes of Northwest Indians. 1996. *Native Peoples of the Northwest: A Traveler's Guide to Land, Art, and Culture.* Seattle: Sasquatch Books.

Hall, S. 1997. "Cultural Identity and Diaspora." In *Undoing Place? A Geographical Reader*, ed. L. McDowell, 231–242. London: Arnold.

———. 1980. "Race, Articulation and Societies Structured in Dominance." In *Sociological Theories: Race and Colonialism*, 305–345, 339. Paris: UNESCO.

———, ed. 1997. *Representation: Cultural Representations and Signifying Practices.* London: Sage Publications, Ltd.

Harjo, J. and G. Bird. 1997. *Reinventing the Enemy's Language: Contemporary Native Women's Writings of North America.* NY: W.W. Norton.

Hau'ofa, E. 2000. "Pasts to Remember." In *Remembrance of Pacific Pasts*, ed. R. Borofsky, 453–471. Honolulu: University of Hawaii Press.

Hechter, M. 1986. "Rational Choice Theory and the Study of Race and Ethnic Relations." In *Theory of Race and Ethnic Relations*, ed. J. Rex, J. Mason, and D. Mason. Cambridge: Cambridge University Press.

———. 1975. *Internal Colonialism: The Celtic Fringe in British National Development, 1536–1966*. Berkeley, CA: University of California Press.

Heininen, L. 2004. "Circumpolar International Relations and Geopolitics." *Arctic Human Development Report*, 207–225. Akureyri, Iceland: Stefansson Arctic Institute.

Hessel, I. 1998. *Inuit Art*. Vancouver/Toronto: Douglas and McIntyre.

Higgins, B. and V. Alia. 1999. "Case Study I: Print Media Coverage Up Here and Outside." In V. Alia, *Un/Covering the North*. Vancouver: UBC Press.

Hijab, N. 2001. *People's Initiatives to use IT for Development*. Background paper for the United Nations Development Programme Human Development Report, 18–19. <http://hdr.undp.org/docs/publications/background-papers/hijab_doc>.

Hoge, W. 2001. "A New, Proud Claim in Norway: 'I Am a Sami.'" *International Herald Tribune*, March 19

Holmes, D. 1989. *Dehcho: Mom, We've been Discovered*. Yellowknife, NWT: Dene Cultural Institute.

Horan, V. 2008. *Global Indigenous TV Network Convenes Inaugural meeting*. Māori Television, June 18.

———. 2008. "Barry Barclay Honoured for Contribution to Indigenous TV Broadcasting." Press release. WITB Māori Television, March 28.

Huhndorf, S. M. 2001. *Going Native: Indians in the American Cultural Imagination*. Ithaca: Cornell University Press.

Hustak, A. 2007. "Gail Guthrie Valaskakis dead at 68." *The Gazette (Montreal)*, July 23 (accessed online).

ICCTA. 2007. *Indigenous Commission Launched to Focus on Information Communications Technologies*. Press release. New York: United Nations, Indigenous Commission for Development of Communications Technologies in the Americas.

IBC. 2006. "Media Release Launch of Nunavut Animation Lab to enhance the sharing of stories by Inuit Artists." October 25.

Indigenous Peoples Summit in Ainu Mosir. 2008. Nibutani Declaration of the 2008 Indigenous Peoples Summit in Ainu Mosir. (online), www.ainumosir2008.com/en/.

Industry Canada. 2006. *Kittiwake Economic Development Corporation to Bring Broadband Internet Service to Additional Aboriginal, Northern and Rural*

Communities. Press Release, March 27. Gander, Newfoundland and Labrador: Industry Canada press release.

International Labour Organisation. 2005. *Convention (No. 169) concerning Indigenous and Tribal Peoples in Independent Countries.* Geneva: United Nations Office of the High Commissioner for Human Rights. <http://www.unhchr.ch/html/menu3/b/62.htm> (accessed 7 September 2008).

Inuit Circumpolar Council (ICC). 2009. *Circumpolar Inuit Launch Declaration on Arctic Sovereignty.* Media Release. Tromsø, Norway: ICC. 28 April.

IRIN. 2009. UN Office for the Coordination of Humanitarian Affairs: IRIN Humanitarian News and Analysis from Africa, Asia, and the Middle East. www.irinnews.org/, May.

ITK. 2005. *Inuit Sign Historic Partnership Agreement.* Press release, May 31.

ITV. 2002. 11:15 pm news broadcast, October 4.

Jaine, L. and D. Taylor, eds. 1995. *Voices: Being Native in Canada.* 2nd ed. Canada: University Extension Press.

Jocks, C. 1996. "'Talk of the Town': Talk Radio." In *Deadlines and Diversity,* ed. Alia, Brennan, and Hoffmaster, 151–172.

Jull, P. 1986. "Greenland's Home Rule and Arctic Sovereignty: A Case Study." In *Sovereignty, Security and the Arctic.* Ontario, Canada: York University.

Kamchatka State University. 2007. Website, www.kamgu.ru/english/.

Keeper, T. 2007. *Harper Government fails Canada with UN Vote.* Website of Tina Keeper, Member of Parliament for Churchill, Manitoba, tinakeeper.liberal.ca/. August 30.

———. 2007. *Conservative Government Must Issue Apology to Residential School Victims.* Website of Tina Keeper, Member of Parliament for Churchill, Manitoba, March 27.

Keeshig-Tobias, L. 1990. "White Lies?" *Saturday Night* magazine, October, 67–68.

Kidd, R. 1997. *The Way We Civilise: Aboriginal Affairs—the untold story.* Brisbane, Australia: University of Queensland Press.

Keith, M.C. 1995. *Signals in the Air: Native Broadcasting in America.* New York: Praeger.

Kleivan, H. 1985. "Contemporary Greenlanders." In *Handbook of North American Indians,* ed. D. Damas. Washington, DC: Smithsonian Institute.

King, J.C.H. and H. Lidchi. 1998. *Imaging the Arctic.* Vancouver: UBC Press.

King, M. 1984. *Maori: A Photographic and Social History*. Auckland: Heinemann.

KNR-online. 2005. June 17. <http://www.knr.gl/English.htm>.

Knudsen, E.R. 2004. *The Circle and the Spiral: A Study of Australian Aboriginal and New Zealand Maori Literature*. Amsterdam/New York: Rodopi.

Kreutz, C. 2009. *Crisscrossed* (blog). http://www.crisscrossed.net/ (accessed 30 April 2009)

Kuhn, A. and K.E. McAllister. 2006. "Locating Memory: Photographic Acts—An Introduction." In A. Kuhn and K.E. McAllister, *Locating Memory: Photographic Acts*, 1–17. New York and Oxford: Berghahn Books.

Kulchyski, P., D. McCaskill, and D. Newhouse, eds. 1999. *In the Words of Elders: Aboriginal Cultures in Transition*. Toronto/London: University of Toronto.

Kuptana, R. 1982. *Inuit Broadcasting Corporation Presentation to the CRTC on Cable Tiering and Universal Pay TV* (speech). Inuit Broadcasting Corporation.

Kymlicka, W. 1995. *Multicultural Citizenship*. Oxford: Clarendon Press.

Lagueux, N. 2006. "The Wapikoni Mobile is visiting the Val-d'Or Native Friendship Centre!" Val-d'Or Native Friendship Centre, *AboriNews.com*, February 6.

Lâm, M.C. 2000. *At the Edge of the State: Indigenous Peoples and Self-Determination*. Ardsley, NY: Transnational Publishers.

Larson, G. 1990. "The Lone Ranger, long since retired, makes an unpleasant discovery" (cartoon). *The Far Side, Yukon News*, October 31, 41.

Lebsock, K. 2006. "Indigenous declaration—Indigenous Peoples' Closing Statement." March 7. New York: American Indian Law Alliance.

Lehtipuu, M. and V. Makela. 1993. *Finland: a travel survival kit*. Australia/US: Lonely Planet.

Lehtola, V-P. 2002. *The Sámi People: Traditions in Transition*. Trans. L. Weber-Müller-Wille. Aanaar (Inari), Finland: Kustannus-Puntsi, Publisher.

———. 2005. "'The Right to One's Own Past.' Sámi Cultural Heritage and Historical Awareness." In *The North Calotte: Perspectives on the Histories and Cultures of Northernmost Europe*, ed. M. Lähteenmäki and P.M. Phlaja. Helsinki, Finland: Publications of the Department of History, University of Helsinki, pp. 83-94.

López-Reyes, R. 1995. "The establishment of a United Nations Permanent Forum of Indigenous Peoples and Autonomous Assembly of Indigenous Peoples." *Indigenous Affairs* 2: 52–56.

Levo-Henriksson, R. 2007. *Media and Ethnic Identity: Hopi Views on Media, Identity, and Communication.* New York and London: Routledge.

Levina, L. 2000. "Women, Society and the Media in North-Eastern Siberia (Russia)." *The Northern Review,* no. 22 (winter): 68–69.

Lincoln, K. 1983. *The Native American Renaissance.* Berkeley and Los Angeles: University of California Press.

Lui, Jr., G.. 1994. "Torres Strait: Towards 2001 (Boyer Lecture)." *Race and Class* 35(4): 11–20.

Lummi Nation. 2006. *Stommish@Sixty* program, Bellingham, WA: Lummi Nation, June 13–24.

Luscombe, R. 2007. "Awake for only 12 days this century—'miracle' of coma woman." *The Guardian,* March 9, 17.

Lutz, E. L. 2007. "Saving America's Endangered Languages." *Cultural Survival Quarterly,* summer: 3.

———. 2007. "Home Stretch." *Cultural Survival Quarterly,* vol. 31, no. 1, spring: 14–23.

Lynge, F. 1998. "Subsistence Values and Ethics." *Indigenous Affairs* 1998(3): 22–25.

———. 1992. *Arctic Wars, Animal Rights, Endangered Peoples.* Hanover and London: University Press of New England.

Lyons, O. 1986. "Spirituality, Equality and Natural Law." In *Pathways to Self-Determination: Canadian Indians and the Canadian State,* ed. Leroy Little Bear, Menno Boldt, and J.A. Long. Toronto: University of Toronto Press.

Maalouf, A. 1996 (trans. 2000). *On Identity.* London: The Harvill Press.

Madslien, J. 2006. "Russia's Sami fight for their lives" (broadcast). *BBC NEWS online.* <http://news.bbc.co.uk/go/pr/fr/-/1/hi/bUSiness/6171701.stm> (accessed 21 December 2006).

Majority World. 2007. Photographs from Darfur, Calcutta, and Kayah state, Myanmar. *Red Pepper,* March, 39.

Manuel, G. and M. Posluns. 1974. *The Fourth World: An Indian Reality.* New York: The Free Press.

Māori Television. 2007a. World Indigenous Television Broadcasting Conference announcement. Māori Television (online), www.maoritelevision.com/.

Māori Television. 2007b. *Be sure to book in early. This is one Hui you will not want to miss* (flyer). World Indigenous Television Broadcasting Conference, unpaginated, November.

Marcus, A. R. 1992. *Out in the Cold: the Legacy of Canada's Inuit Relocation Experiment in the High Arctic.* Copenhagen: International Work Group for Indigenous Affairs.

Marcus, G.E. and D. Cushman. 1982. "Ethnographies as Texts." *Annual Review of Anthropology,* vol. 2: 25–69.

Marks, K. 2008. "The big question: After three months in power, how has Kevin Rudd changed Australia?" *The Independent,* February 20, 33.

———. 2004. "Thousands of Maoris march to defend 'their' beaches." *The Independent,* May 6. <http://www.independent.co.uk>.

Marshall, E. Z. 2006. *Indigenous Media Outlets: Web Based Research by Emily Marshall for Valerie Alia.* April.

———. 2005–2006. Research work for Valerie Alia. Leeds, UK: Leeds Metropolitan University.

Marshall, P., director. 1990. *Awakenings* (film).

Masterson, A. 2003. "More than once were warriors out of their tiny minds." *The Age,* January 21. <http://www.theage.com.au>.

Mattes, C. and S.F. Racette. 2001. Rielisms, Winnipeg: Winnipeg Art Gallery.

McMaster, G. 1998. *Reservation X: The Power of Place in Contemporary Aboriginal Art.* Seattle: University of Washington Press.

McMaster, G and L-A. Martin, eds. 1992. *Indigena: Contemporary Native Perspectives.* Vancouver: Douglas & McIntyre.

McFarlane, P. 1993. *Brotherhood to Nationhood: George Manuel and the Making of the Modern Indian Movement.* Toronto: Between the Lines.

McQuire, A. 2007. "We know Howard and Costello, but do we know Rudd?" Special Election '07 Edition, *National Indigenous Times,* November 24.

Meadows, M. 2001. *Voices in the Wilderness: Images of Aboriginal People in Australian Media.* Westport, CT and London: Greenwood Press.

———. 1996. "Ideas from the Bush: Indigenous Television in Australia and Canada." *Canadian Journal of Communications.*

Meadows, M and H. Molnar. 2002. "Bridging the Gaps: towards a history of Indigenous media in Australia." *Media History,* vol. 8, no. 1: 9–20.

Medicine, B. 2001. *Learning to be an Anthropologist and Remaining "Native."* Urbana / Chicago: University of Illinois Press.

Minogue, S. 2006. "Iqaluit to become a hotspot." *Canadian Press,* April 19.

Miller, C. 2003. *Canadian Broadcast Timeline.* Canadian Radio History Page. <http://www.odxa.on.ca/archives/timelinebc.html>.

Minde, H. 1995. "The International Movement of Indigenous Peoples: an Historical Perspective." In *Becoming Visible—Indigenous Politics and Self-government*, ed. T. Brantenberg, J. Hansen, and H. Minde. Tromsø, Norway: University of Tromsø, Sámi dutkamiid guovddás—Centre for Sámi Studies.

Minority Rights Group. 1994. *Polar Peoples: Self-Determination and Development*. London: Minority Rights Publications.

Mitchell, A. 2001. "How the North is getting burned." *The Globe and Mail*, June 5.

Molnar, H. and M. Meadows. 2001. *Songlines to Satellites: Indigenous Communication in Australia, the South Pacific and Canada*. Annandale, New South Wales, Australia: Pluto Press.

Morgan, L. 1988. *Art and Eskimo Power: The Life and Times of Alaskan Howard Rock*. Fairbanks: University of Alaska Press.

Morilova, A.V. 2007. "Aborigen Kamchatki" (online), www.usu.ru/ip/block.php?sid+5&lng+en.

Mott Foundation. 2005. "ABC Ulwazi uses listeners' associations to link radio with citizen education, action." *Charles Stewart Mott Foundation*. <http://pubs.mott.org/news/online.asp?c+al> (accessed 7 September 2008).

Moyers, B. 2007. *Life on the Plantation*. Address to the National Conference for Media, Memphis, Tennessee. Posted on *truthout* website, www.truthout.org/, January 12.

Mufune, J. 2001. "Broadcasting in Zimbabwe." <http://www.niza.nl/uk/campaigns/media/08/index.html>.

Murphy, L. 2002. "Australian Democracy and Aboriginality." Ph.D. Work in Progress Paper, School of Political Science and International Studies. Brisbane, Australia: University of Queensland.

———. 2000. "Who's Afraid of the Dark? Australia's Administration in Aboriginal Affairs." MA Thesis, Center of Public Administration. Brisbane, Australia: University of Queensland.

Murphy, L. and M. Briggs. 2003. "The Sad Predictability of Indigenous Affairs." *Arena Magazine*, October 1: 1–3.

Mutuku, T. 2005. "Fiji Community Radio Initative with Suitcase Radio." WACC Pacific Blog, January 17. <http://www.waccglobal.org/de/regions/pacific/pacific_news> (accessed 7 September 2008).

Nanaimo Bulletin. 2009. "Dance unites spirits." *Nanaimo Bulletin*, May 4.National Indigenous Media Association of Australia (NIMAA).

1998. NIMAA: "The Voice of Our People." Fortitude Valley, Queensland, Australia: NIMAA.

National Post. 2007. "Gail Valaskakis, R.I.P." July 21 (online).

Native American Times. <http://www.nativetimes.com/index.asp?action=displayarticle&article_id=8286>.

Negi, M., Ki Awaaz, M. 2007. "Local self help groups strengthen community radio." *CI News*, April 24. New Delhi: UNESCO.

New Zealand Ministry of Foreign Affairs and Trade. 2004. He kāp rero mā aAotearoa / about New Zealand Aotearoa. The New Zealand Ministry of Foreign Affairs and Trade.

Nietschmann, B. 1994. "The Fourth World: Nations versus States." In *Reordering the World: Geopolitical Perspectives on the 21st Century*, ed. G. D. Demko and W. B. Wood, 225–242. Boulder, CO: Westview Press.

NIMAA. 2005. "Indigenous Media Codes of Ethics (1st Draft)" in press pack, NIMAA: "The Voice of our People." Fortitude Valley, Queensland, Australia: National Indigenous Media Association of Australia.

NITV. 2007. "Black TV Launches on Black Friday." Online Media Release, nitv.org.au/, July 13.

Norimitsu, O. 2008. "Recognition for a People Who Faded as Japan Grew."

New York Times Global Policy Forum, July 3. <http://www.globalpolicy.org/nations/sovereign/sover/erg/2008/0703ainu.htm> (accessed 12 September 2008).

Northern Perspectives. 1995. "Aboriginal Rights and Land Claims in New Zealand." Special Issue, 23 (2), Summer.

Norway. 2007. 'sámi history and policy." *Society and Policy, Norway, the official site in the UK* [online], http://www.norway.org.uk/facts/sami/policy/policy.htm [6 September 2008]

Nunatsiaq News. 2001. "Indigenous Peoples show support for Thule Inughuit. Iqaluit." February 8.

Nunatsiaq News. 2002. "Chukotka's governor supports Inuit hunting." May 9, p. 2.

Nuttall, M. 1998. *Protecting the Arctic: Indigenous Peoples and Cultural Survival*. Oxford: Taylor and Francis.

———. "Greenland: Emergence of an Inuit Homeland." In *Polar Peoples: Self-determination and Development*, ed. Minority Rights Group, 1–28. London: Minority Rights Publications.

NWIN. 2006. *NorthWest Indian News*, Program 14, March. "News by, for and about Indian People of the Northwest." Program produced by the

Tulalip Tribes Communications Department in Tulalip, WA and distributed by DVD to Native American/First Nations across the northwestern United States.

O'Brien, H. 2007. "The complete guide to Lapland." *The Independent*, August 4, 14.

O'Hagan, J. 2002. "'The Power and the Passion'": Civilizational Identity and Alterity in the wake of September 11." In *Identity and International Relations: Beyond the First Wave*, ed. K. Dunn and P. Goff.

O'Neill, J. and T. Dalrymple. 2008. "Native leaders hail apology as a historic turning point." *Times Colonist*, June 12, A12.

Okalik, P. 2001. *What Does Indigenous Self-Government Mean?* Speaking notes, Brisbane, Australia: Brisbane Dialogue. Government of Australia, Parliament website, http://parlinfo.aph.gov.au/parlInfo/. August 13.

Olwig, K.F. and K. Hastrup. 1997. "Introduction." In *Siting Culture: The Shifting Anthropological Object*, ed. K.F. Olwig and K. Hastrup, 1–14. London: Routledge.

Paine, R. 1977. "The Path of Welfare Colonialism." In *The White Arctic: Anthropological Essays on Tutelage and Ethnicity*, ed. R. Paine, 7–28. Toronto: University of Toronto Press.

Pearsall, J. and B. Trumble, eds. 1996. *The Oxford English Reference Dictionary*. 2nd ed. Oxford: Oxford University Press.

Pennanen, J. 1998. *Siida: The Sámi Museum Guide*. Inari, Finland: Siida Sámi Museum and Northern Lapland Visitor Centre.

Perrot, M. 1993. "L"état des médias en Tchoukotka." In *Sibérie III: Questions sibériennes*, ed. M. Perrot. Paris: Unpublished monograph.

Perry, R. 1996. *From Time Immemorial: Indigenous Peoples and State Systems*. Austin: University of Texas.

Petersen, R. 1984. "The Pan-Eskimo Movement." In D. Damas, *Arctic*, 724–728. Washington, DC: Smithsonian Institution.

Petrone, P. ed. 1992. *Native Literature in Canada: from the oral tradition to the present*. Oxford: Oxford University Press.

———. ed. 1988. *Northern Voices: Inuit Writing in English*. Toronto: University of Toronto Press.

———. ed. 1983. *First People First Voices*. Toronto: University of Toronto Press.

Piercy, M. 1980. *The Moon is Always Female*. In B. Moyers, *Life on the Plantation*.

Pietikäinen, S. 2003. "Indigenous identity in print: representations of the Sami in news discourse." *discourse & Society*, vol. 14, no. 5: 581–609.

Pilkington, D. (Traditional name, N. Garimara). 2002. *Follow the Rabbit-Proof Fence*. Brisbane: University of Queensland Press

Pinter, H. 1982. *A Kind of Alaska* (play).

Pitseolak, P. 1977. *Peter Pitseolak's Escape from Death*. Ed. D. Eber. Toronto: McClelland and Stewart.

Pitseolak, P. and D. Harley Eber. 1975. *People from Our Side: A Life Story with Photographs and Oral Biography*. Translated by A. Hanson. Edmonton: Hurtig Publishers. (Reprinted 1993, with a new preface by D. Harley Eber, Montreal and Kingston: McGill-Queen's University Press.)

Plunkett, J. 2007. "Radio days are here again as Britons tune, click and plug into digital age." *The Guardian*, August 17, 3.

———. 2007. "Go figure: Radio listening." *The Guardian*, April 30, 9.

Podur, J. 2005. "Colombian Government's Counteroffensive: Targeting the indigenous movement." *Znet*, May 20. <http://www.zmag.org/>.

Portalewska, A. 2005. "Radio tower in El Tablon, Solola." Guatemala Indigenous Community Radio Project Cultural Survival website: AboriNews.com, October 3.

Power Radio. 2006. *Seizing the Airwaves: a free radio handbook*.

Pratt, M.L.. 1992. *Imperial Eyes: Travel Writing and Transculturation*. London: Routledge.

Qitsualik, R. A. 2003. "Playing Cowboys and Inuit." *Indian Country Today*. Website, www.indiancountrytoday.com/, January 23.

Quarterman, J.S. 1991. "Telecomputing in the New Global Networks." In *Communications in History: Technology, Culture, Society*, ed. D. Crowley and P. Heyer, 341–348. White Plains, New York: Longman.

Quigley, M.. 1982. "Language of Conquest, Language of Survival." *Canadian Forum*, November, 14–15.

Radio Hele Norge. 1999. *1999 Annual Report*. Website, www.p4.no/ (accessed 2006).

Radio and TV Greenland. 2001. *News and Media: Radio and TV Greenland*. <http://www.randburg.com>.

Red Pepper. 2007. "Peoples of the Majority World—telling their own Story!" *Red Pepper*, March, 39.

Reitz, J. G. 1980. *The Survival of Ethnic Groups*. Toronto: McGraw-Hill Ryerson.

Richie, D. 1996. "Dersu Uzala." *The Films of Akira Kurosawa*, 195–203. Berkeley/Los Angeles: University of California.

Ritter, R.M. ed. and compiler. 2000. *The Oxford Dictionary for Writers and Editors.* 2nd ed. 172. Oxford: Oxford University Press.

Rizvi, H. 2007. "Journalist Deaths Still Climbing Every Year." *Inter Press Service News Agency (IPS),* December 18.

Robie, D. 2005. "'Four Worlds' news values: Media in transition in the South Pacific." Journalism and the Public (Journalism Education Association) conference, Griffith University, Gold Coast, Australia, November 29–December 2.

Roth, L. 2006. *Something new in the air: the story of first peoples television broadcasting in Canada.* Montreal, Kingston (Ontario, Canada), London, Ithaca, New York: McGill-Queen's University Press.

———. 2000. "Bypassing of Borders and Building of Bridges: Steps in the Construction of the Aboriginal Peoples Television Network in Canada." *Gazette* vol. 62(3-4): 251–269. Sage Publications.

———. 1996. "The Politics and Ethics of Inclusion: Cultural and Racial Diversity in Canadian Broadcast Journalism." In *Deadlines and Diversity,* ed. Alia, Brennan and Hoffmaster.

———. 1995. "(De)romancing the North." *Border/Lines 36:* 36–43.

———. 1993. "Mohawk Airwaves and Cultural Challenges: Some Reflections on the Politics of Recognition and Cultural Appropriation after the Summer of 1990." *Canadian Journal of Communication,* vol. 18, no. 3: 315–331.)

Roth, L. and G. Guthrie Valaskakis. 1989. "Aboriginal Broadcasting in Canada: A Case Study in Democratization." In *Communication for and against democracy,* ed. M. Raboy and P. A. Bruck. Montreal: Black Rose.

The Royal Ontario Museum. 2006. "Gallery of Canada: First Peoples The ROM's First Peoples collection featured in main floor gallery, opening Dec. 26, 2005." January 19. Toronto: Royal Ontario Museum (ROM).

Ryan, A. J. 1999. *The Trickster Shift: Humour and Irony in Contemporary Native Art.* Vancouver: UBC Press.

Ryan, A. J. 1991. *The cowboy/Indian Show: Recent Work by Gerald McMaster.* Kleinberg, Ontario: McMichael Canadian Art Collection.

Sacks, O. 1983 (1973). *Awakenings.* Rev. ed. New York: Dutton.

Sahlins, M. 1999. "What is anthropological enlightenment/ some lessons from the twentieth century." *Annual Review of Anthropology* 28: i–xxiii.

Said, E. 1993. *Culture and Imperialism.* New York: Vintage.

———. 1986. "An Ideology of difference." In *'Race.' Writing, and Difference*, edited by L.H. Gates, Jr., 38–58. Chicago: University of Chicago Press.

Salazar, J. F. 2003. "Articulating an activist imaginary: Internet as counter public sphere in the Mapuche movement, 1997-2000." In *Media International Australia incorporating Culture and Policy*, No.107, pp. 19-30. Brisbane: University of Queensland.

Salazar, J. F. and A. Córdova. 2008. "Imperfect Media and the Poetics of Indigenous Video in Latin America. In *Global Indigenous Media*, 39-57.

Salleh, A. 2005. "News in Science—Earliest human footprints in Australia." *ABC Science Online*, Australian Broadcasting Corporation, December 21.

"Sámi history and policy." 2007. *Society and Policy, Norway, the official site in the UK*. <http://www.norway.org.uk/facts/sami/policy/policy.htm> (accessed 6 September 2008).

Sami Parliament. 2002. "The Sami in Finland." Inari, Finland: Sami Parliament Secretariat. Publications by Sami Parliament.

Samefolket. 2006. Ostersund, Sweden: *Samefolket* English version, online edition, info.samefolket.se/.

Schlein, L. 2007. "UN rights body adopts indigenous rights declaration despite Canada's No vote." *Canadian Press* (online), June 29.

Schubert, M. 2007. "Rudd gets to it, with vows and vision."

Sciadas, G., ed. 2005. *From the Digital Divide to Digital Opportunities: Measuring Infostates for Development*. Montreal, Canada: Claude-Yves Charron, International Telecommunication Union and Orbicon International Secretariat of UNESCO. <http://www.digitaldivide.net/articles/view.php?ArticleID=123>.

Shnukal, A. 2001. "Torres Strait Islanders." In *Multicultural Queensland 2001: 100 years, 100 communities, a century of contributors*, ed. Maximilian Brändle. Brisbane: Multicultural Affairs Queensland, Queensland Government.

Silis, I., ed. 1996. "Greenland on its Way—Home Rule since 1979." Copenhagen, Ministry Danida, Greenlands Television.

———. 1982. *Greenland Today*. Willimantic: Curbstone Press.

Singe, J. 1989. The Torres Strait: People and History. Brisbane, Australia: University of Queensland Press.

Sinclair, J. and S. Cunningham. 2000. "Diasporas and the Media." In *Floating Lives: The Media and Asian Diasporas,* ed. S. Cunningham and J. Sinclair, 1–34. Australia: University of Queensland Press.

Singh, S. et al., eds. 2001. *Aboriginal Australia and the Torres Strait Islands.* Australia: Lonely Planet.

Sjödén, G. 2003. *Spring 2003: The Four Elements, Part One.* Stockholm: Gudrun Sjödén.

Slezkine, Y. 1997. *Arctic Mirrors: Russia and the Small Peoples of the North.* Ithaca, New York: Cornell University Press.

Smith, B.L. and M.I. Cornette. 1998. "Electronic Smoke Signals: Native American Radio in the United States." *Cultural Survival Quarterly,* summer: 28–31.

Smith, L. Tuhiwai. 1999. *Decolonizing Methodologies: Research and Indigenous Peoples.* London and New York: Zed Books.

Solbakk, J.T. 2006. *The Sámi People—A Handbook.* Karasjok, Norway: Davvi Girji OS.

———. 1997. "Sámi Mass Media: Their Role in a Minority Society." In Gaski, H. *Sámi Culture in a New Era,* 172–220.

Solomon, A. 1990. "War Song, There is no middle ground." In Solomon, A. (1990). *Songs for the people: Teachings on the natural way.* Toronto: NC Press.

Sperschneider, W. 1998. *100 Years of Greenland on Film.* Unpublished Ph.D. diss. and documentary film, Arhās, Denmark: Arhās University.

Spivak, Gayatri Chakravorty. 1995. "Subaltern Studies: Deconstructing Historiography." *The Spivak Reader: Selected Works of Gayatri Spivak,* ed. Donna Landry and Gerald MacLean. NY: Routledge.

———. 1990. *The post-colonial critic: Interviews, strategies, dialogues.* New York: Routledge.

Staples, B. 2003. "Racial discrimination and black Indians." *International Herald Tribune,* September 17, p. 9.

Stavenhagen, R. 2001. *Report on the Situation on Human Rights and Fundamental Freedoms of Indigenous People.* United Nations.

Steffens, L. 1931. *The Autobiography of Lincoln Steffens.* New York: Harcourt Brace and Company.

Survival International. 2008. "Bruce Parry Brings Together Music World's Brightest Starts for 'Survival' Album." *Survival International Press Release,* September 15.

———. 2006. "UN Declaration on Indigenous People Blocked." Press Release, November 30.

———. 2006. "Indigenous Peoples Win Historic Vote on Rights—Canada Votes Against." Online press release, August 7.

———. 2006. "UK: BBC launches second series of hit TV show 'Tribe.'" Press release, July 14.

Suzuki. 2002. *Room for the tribe* (advertisement). Japan: Suzuki.

Svensson, T.G. 1985. "The Sami and the Nation State: Some Comments on the Ethnopolitics in the Northern Fourth World." *Études/Inuit Studies*, VIII, 2, pp. 158-166.

Swaney, D. (ed.) 1999. *The Arctic*, London: Lonely Planet.

Svensson, T.G. 1985. "The Sami and the Nation State: Some Comments on the Ethnopolitics in the Northern Fourth World." *Études/Inuit/Studies*, vol. 8, no. 2: 158–166.

Swiercinsky, D.P., T.L. Price, and L.E. Leaf. 1993. *Traumatic Head Injury*. Kansas City, KS: The Head Injury Association of Kansas and Greater Kansas City, Inc.

Sykes, J.B., ed. 1982. *The Concise Oxford English Dictionary of Current English*. 7th ed. Oxford: Oxford University Press.

Taksami, C. 1990. "Opening Speech at the Congress of Small Indigenous Peoples of the Soviet North." *Indigenous Peoples of the Soviet North*, IWGIA Document No. 67, Copenhagen: International Workgroup for Indigenous Affairs.

Tamahori, L 1994. *Once Were Warriors*. Film, New Zealand.

Taylor, L.A. 2009. In "Overview." *The State of the News Media: An Annual Report on American Journalism*, Washington, DC: PEW Research Center for the People and the Press (online), www.pewcenter.org/.

Te Araratuku Whakaata Irirangi Maori. 2003. Māori Television Service (Te Araratuku Whakaata Irirangi Maori) Act 2003, Section 8.

The Telegraph. 2007. *Telegraph.co.uk online*, November 27.

Temman, M. 2006. "The Ainu Will Be Voiceless in Japan." *Libération* [*Truthout online*], May 10. <http://www.truthout.org/article/the-ainu-will-be-voiceless-japan> (accessed 7 September 2008).

Tester, F. J. and P. Kulchyski. 1994. *Tammarniit (Mistakes): Inuit Relocation in the Eastern Arctic 1939–63*. Vancouver: UBC Press.

Thérrien, R. 1980. *The 1980s: a Decade of Diversity—Broadcasting, Satellites, and Pay-TV. Report from the Committee on Extension of Service to Northern and Remote Communities (The Thérrien Report)*. Ottawa: Minister of Supply and Services Canada (CRTC), Catalogue No. BC 92-24/1980E.

Thompson, F. 1992. "British newspaper article deserves 'harsh rebuttal'—Holman Mayor." *News/North*, November 2, A3, A30.

Tompkins, J. 1986. "'Indians': Textualism, Morality, and the Problem of History." In Gates, *"Race," Writing, and Difference.*

Travelmood. 2007. *Experience Australia's Cultural Heart* (advertising campaign). Australia: Travelmood.

Turner, G. 1993. *Nation Culture Text: Australian cultural and media studies.* London: Routledge.

Turner, T. 2002. "Representation, Politics, and Cultural Imagination in Indigenous Video." In *Media Worlds: Anthropology on New Terrain*, ed. F. D. Ginsburg, L. Abu-Lughod, and B. Larkin, 75–89. Berkeley/Los Angeles/London: University of California Press.

Turpel, M. E. 1989/90. "Aboriginal Peoples and the Canadian Charter: Interpretive Monopolies, Cultural Differences." *Canadian Human Rights Yearbook*, 6: 3–45.

Twigg, A. 2005. *Aboriginality: The Literary Origins of British Columbia.* Vol. 2. Vancouver: Ronsdale Press.

Twohy, P. J. 1999. *Beginnings: A Meditation on Coast Salish Lifeways.* La Conner, WA: Swinomish Spiritual Center at St. Pauls Mission.

Uden, M. and A. Doria. 2006. Sámi Network Connectivity Project website, www.snc.sapmi.net/.

UNESCO. 2007. "Telecentre on Wheels: A new way to access information in rural India." *CI News*, December 18. New Delhi: UNESCO.

———. 2007. "Advancing gender equality in the Pacific." *CI News*, December 6. Apia, Samoa: UNESCO.

———. 2007. "Community Radio Rides a Popularity Wave." *CI News*, February 15. New Delhi: UNESCO.

———. 2006. "Tripura tribal community in Bangladesh finds a voice." *CI News*, December 11. New Delhi: UNESCO.

———. 2007. "World Press Freedom Day." *CI News*, May 14. Apia, Samoa: UNESCO.

———. 2007. "India to establish 4000 community radio stations under new community radio policy." *CI news*, March 15. New Delhi: UNESCO.

———. 2006. "Voice from the Pacific Grassroots: pactoc.telecentre.org." *CI News*, February 28. New Zealand National Commission for UNESCO.

———. 2006. "Community Multimedia Centres." *CI News*, February 10.

———. 2006. "Voice from the Pacific Grassroots: pactoc.telecentre.org." February 28. New Zealand National Commission for UNESCO.

———. 2006. "Ta-Oi District Radio hits the airwaves." *News Release*, July 2. New Delhi: UNESCO Liaison Office.

United Nations. 2006. *United Nations Declaration on the Rights of Indigenous Peoples*, Resolution 2006/2, United Nations Human Rights Council, June 29.

University of Western Ontario. 2007. *First Nations in the News Media*. Faculty of Information and Media Studies, Media, Information and Technoculture, online, http://www.fims.uwo.ca/.

Valaskakis, G. Guthrie. 1995. "Sacajawea and Her Sisters: Images and Indians." In M. Burgess and G. Guthrie Valaskakis, *Indian Princesses and Cowgirls: Stereotypes from the Frontier*, 11–39. Montreal: OBORO.

———. 1993. "Dance Me Inside: Pow Wow and Being 'Indian.'" *Fuse* 16(5/6), summer: 39–44.

———. 1988. "The Chippewa and the Other: Living the Heritage of Lac du Flambeau." *Cultural Studies*, 2(3), October.

———. 1982. "Communication and Control in the Canadian North—The Potential of Interactive Satellites." *Etudes/Inuit Studies* 6(1): 19–28.

Verán, C. 2006. "AlterNATIVE Media: Indigenous Video Activists Set the Scene to Be Heard." *Fellowship* (online) Fellowship of Reconciliation. May/June.

Vitebsky, P. 2001. Oral seminar presentation and discussions, Cambridge, UK: Scott Polar Research Institute, University of Cambridge (unpaginated).

Vørren, O. 1994. *Saami, Reindeer, and Gold in Alaska: The Emigration of Saami from Norway to Alaska*. Illinois: Waveland.

Waite, M., ed. 1996. *The Oxford Colour Spelling Dictionary*. Oxford: Clarendon Press.

Walsh, A. 2006. "Re-placing History: Critiquing the Colonial Gaze through Photographic Works by Jeffrey Thomas and Greg Staats." In *Locating Memory: Photographic Acts*, ed. Kuhn and McAllister, 21–51.

Wasserman, H. and S. Jacobs. 2003. *Shifting Selves: Post-apartheid Essays on Mass Media, Culture and Identity*. Kwela, South Africa: Kaapstad.

Webster, S. 1998. *Patrons of Maori Culture: Power, Theory and Ideology in the Maori Renaissance*. Dunedin: University of Otago Press.

Webster's Online Dictionary. 2008. "Outlaw." <http://www.websters-online-dictionary.org/definition/outlaw> (accessed 5 September 2008).

Weekly Indigenous News. 2005. "Taiwan launches Indigenous Television Network." July 15.

Wessendorf, K. 2005. "Introduction." In *An Indigenous Parliament? Realities and Perspectives in Russia and the Circumpolar North*, ed. K. Wessendorf. Document No. 116, 8–21. Copenhagen: IWGIA (International Work Group on Indigenous Affairs).

White Eye, B. (E.). 1996. "Journalism and First Nations." In *Deadlines and Diversity*, ed. Alia, Brennan and Hoffmaster.

Williams, J., C. Niven, and P. Turner. 2000. *New Zealand*. 10th ed. London: Lonely Planet.

Wilmer, F. 1993. *The Indigenous Voice in World Politics: Since Time Immemorial*. London: Sage Publications.

Wilson, P. and M. Stewart, eds. 2008. *Global Indigenous Media: Cultures, Poetics, and Politics*. Durham, North Carolina: Duke University Press.

WITBN 2009. *WITBN Newsletter* Issue 14, 20 February (online) World Indigenous Television Broadcasters Network (WITBN).

Woodward, K., ed. 1997. *Identity and Difference*. London: Sage/the Open University.

Wallheimer, B. 2006. *Norwich [Connecticut] Bulletin*, February 23.

Yamamoto, K. 1993. "Foreword." *The Committee to Organize the Exhibition of Ainu Pictures to Promote Human Rights. Exhibition of Ainu Pictures to Promote Human Rights* (catalogue), October 10, 5–6. Sapporo city, Japan: Hatamoto printing Co.

Yasumoto, M. 2007. "Ainu 'rebels' mix it up to get message across." Japan *Times online*, November 22.

YLE Radio Finland. 2007. Inari, Finland: YLE Sámi Radio website.

———. 2002. *Radiosi maailmalla; Finländsk världsradio* (leaflet). YLE Radio Finland, 28 October 2001–31 March 2002.

Zakharov, D.N. 2005. "The Example of Sakha (Yakutia)." In *An Indigenous Parliament?*, ed. K. Wessendorf, 88–97.

Zakoda, E. 2006. "The Asahi Shimbun. Voice of the Ainu speaks to the people." *International Herald Tribune*, May 22. <http://www.iht.com> (accessed 24 May 2006).

Additional References and Sources

Alia, V. 2008. Email to Miriam Ross, public relations, Survival International. September 15, 19:12.

———. 2006. Personal communication; Sámi Radio headquarters. Karasjok, Norway. August.

———. 2006. Personal communication, Alanis Obomsawin. Durham, UK: British Association of Canadian Studies conference.

———. 2006 interviews.

———. 2000–2001. Interviews conducted for research project funded by the Canadian High Commission.

Brewis, A. 2009. Personal Communication (discussion with V. Alia). April.

Bristow, G, R. Condon, and R. Kuptana. 1992. Unpublished Letter to the Editor of the *Telegraph Magazine*, sent from Holman Island, Canada.

George, C. 1996. Personal communication, conversation with Valerie Alia at his home at Stoney First Nation, Ontario, Canada.

Hadl, G. 2008. Personal communications, 20 April 2008; 30 August 2008.

Helander, K. R. 2006. Personal communication, 15 December.

Posluns, M. 2007. E-mail discussion with V. Alia, November 27–29.

Robertson, C. (2003) Personal communication.

Ross, M. 2008. E-mail to Valerie Alia [valerie_alia@hotmail.com], "Bruce Parry Brings Together Music World's Brightest Stars for 'survival" Album, Survival International, [mr@survival-international.org],Tuesday 16 September 18:13.

Smoke, D. 2006. Personal communication, Toronto May 30.

Steffens, P. 2007. Personal communication and discussions, February.

———. 1999. Personal communication (1998) quoted in Alia, *Un/Covering the North*, 197.

Index

B

G

H

I

J

K

L

N

S

T

Y

Z

LaVergne, TN USA
14 January 2011
212480LV00001B/34/P

9 781845 454203